Peter Bongaarts

Algebraic Dynamical Systems

A general framework for the description of physical systems

Second corrected edition

MINKOWSKI
Institute Press

Peter Bongaarts
Institute Lorentz for Theoretical Physics
Niels Bohrweg 2, Leiden, NL-2333 CA
The Netherlands

Cover: Image taken from https://recordnotfound.com/autograd-HIPS-18214

ISBN: 978-1-927763-52-0 (softcover)
ISBN: 978-1-927763-53-7 (ebook)

Minkowski Institute Press
Montreal, Quebec, Canada
http://minkowskiinstitute.org/mip/

For information on all Minkowski Institute Press publications visit our website at http://minkowskiinstitute.org/mip/books/

"An algebraic theory of physics is affected with just the inverted advantages and weaknesses [of prevailing ideas], ... It would be especially difficult to derive something like a spatio-temporal quasi-order from such a schema. But I hold it entirely possible that the development will lead there"

(Albert Einstein in a letter to the physicist H.S. Joachim)

Ludwig Faddeev, dean of Russian mathematical Physics
(1934-2017)

This picture was taken by the author at the occasion of his visit to Leiden
and reproduced here with the permission of the late Professor Faddeev

Voor Piek

Contents

PREFACE

Historical remarks

The idea of an algebraic formulation of quantum mechanics found its first clear expression in a fundamental paper by Irving Segal [1], published in 1947. For a long time this paper went unnoticed, but finally, some twenty years later, Rudolf Haag and Daniel Kastler took up its ideas and applied it to quantum field theory, then, as now, a very important and interesting, but also, mathematically speaking, a very problematic topic. This resulted in their 1964 paper [2]. The subject was developed further by – among many others – Huzihuro Araki, Hans-Jürgen Borchers, Detlev Buchholz and Klaus Fredenhagen; its final form is now called *algebraic quantum field theory*.

This could be called the end point, at least as far as the algebraic approach to quantum theory goes. The next step, the idea that *all* physical theories, both quantum and classical, can be described within a single algebraic framework, with non-commutative algebras for the quantum case, and commutative algebras for the classical situation, is something that many mathematical physicists have probably been aware of – in a vague way, but was nevertheless rarely explicitly discussed or expressed.

To the knowledge of this writer there are only two books that do this, the book by Ludwig Faddeev and O.A. Yakubowskiĭ [3], and that by Franco Strocchi [4]. The first is an elegant but rather short book, which does not enter into details, the second one admits as algebras only C^*-algebras, which is much too restricted.

In my recent book "Quantum Theory. A Mathematical Approach" [5], this general algebraic point of view is one of the main underlying themes. In this short book it will be discussed in a more extensive and explicit manner.

Theoretical physicists strive for unification of theories; general formalisms have always had a strong appeal. A matter of aesthetics, a wish for elegance. These were and still are important driving forces in the development of physics and mathematics. (Remember, however, as a warning, a quote by Ludwig Boltzmann, one of the fathers of statistical mechanics: "Eleganz ist für Schneider und Schuster" ("Elegance is for tailors and shoemakers")).

It has also to be admitted that a general theory may not always be helpful in discussing explicit examples. This limits its use in practice, even though it does not diminish its general attractiveness and importance.

However, in the case of algebraic dynamical systems there is an important application, namely to the study of the foundations of quantum theory. By means of this formalism this author has been able to bring to light surprising but also rather provocative facts, such as non-existence or irrelevance of the so-called 'collapse of the wave function', or that the famous 'Schrödinger Cat' thought experiment describes just a simple classical stochastic process, which does not say anything on quantum mechanics. See [6].

Possible readership of this book

This short book is meant for physics and mathematics students, third year and higher, and of course also for physicists and mathematicians generally. Less mathematical and physics background is needed than for my earlier book [5]. Nevertheless quite a few mathematical, respectively physics notions will appear that are unfamiliar for physics, respectively mathematics students, given the almost complete separation between the teaching of mathematics and physics at present. Long explanations, with precise details and proofs, as I gave in my earlier book, are out of the question here, given the size of this book.

There is a solution for this problem. Whenever a mathematical term appears that may raise questions for some of the students, a 'Mathematical intermezzo' or 'Physics intermezzo' is given, containing a short sketch of the main properties of the notion in question, sometimes without, for instance, complete sets of axioms or theorems with their proof. If a student finds this satisfactory, he can go on reading. Otherwise he may consult the references added to the intermezzo. Most of these are available on internet; the links are given in list of references at the end of each chapter, which makes instant consultation possible. Some topics are so important that they need more details. For this we give more extended intermezzos, denoted as 'Mathematical / Physics intermezzo - extra'.

Readers who are in possession of the e-version of this book get direct access to the references that have an internet link, as clicking on such a link in the e-book file gives immediately the article or book in question. For owners of the hard copy of the book looking at these books or articles is more complicated, as they have to type the internet address – sometimes quite long and complicated - in their browser. For this the author has prepared an e-mail message with all the references with internet links. The recipients of this e-mail can obtain the books and articles by clicking on the links in this e-mail and obtain them in the same manner as in the case of the e-book.

This e-mail message can be requested from

`p.j.m.bongaarts[at]xs4all.nl`

Finally, for a complete and rigorous treatment of the mathematics in this book the 'supplementary chapters' of my earlier book [5] will be useful.

There are lists of these intermezzos at the end of the book, with the chapters and sections where they occur, one alphabetical, the second one in order of appearance in the text. This makes this collection into a useful and convenient dictionary. The subject index at the end of the book will also be helpful in this respect.

This book is not meant to take the place of a course or book on whatever specific topic in physics; instead, it is an addition to the standard curriculum, and tries to encourage a broader view on correspondence and general structure of mathematics and physics, not a luxury in this era of specialization in the sciences.

The general theoretical scheme presented in this book is far from complete, with many details missing. At various places remarks will be made on things that so far are only known in the form of sketches.

An important topic is in this respect the theory and application of locally convex algebras, more general than C^*- or von Neumann algebras. Much is known about such algebras in the mathematical literature, but they have scarcely been used in mathematical physics. In Chapter 8 we shall argue that these algebras are very useful in the elegant and natural Uhlmann-Borchers algebraic formulation of Wightman axiomatic field theory. They cannot be avoided in the case of the quantization of the Maxwell field, as we shall show in Chapter 9. This means that in the formulation of algebraic axiomatic field theory in Chapter 10 these more general algebras have also to be included.

Historically the Maxwell field was the first field to be quantized. As such it became part of quantum electrodynamics, the theory that describes the interaction between photons, electrons and positrons, and which is, in its 'renormalized' form, up till now the most accurate quantum field theory, that has been tested experimentally. It is therefore strange that the Maxwell field is not mentioned in most standard textbooks on Wightman theory or algebraic quantum field theory.

Maybe the reason is that mathematical physicists shy away from this subject because it leads to 'pseudo-Hilbert spaces', i.e. spaces with a non-degenerate but indefinite inner product, spaces which are believed to be unphysical, and their use contrary to the principles of quantum theory. This is a misunderstanding, as will be explained.

Most of the examples given in the book are well-known. We shall however also present some interesting recent topics. One example is the work by Frank den Hollander c.s. on what is usually called "Stochastic graph theory", because it is classical statistical mechanics based on graphs [7]. It is by now quite popular and has a wide range of interesting applications outside physics. There are possible non-commutative generalizations, still only partly explored. This is discussed in Chapter 11.

A second example, mentioned already above, is the Maxwell quantum field, which has a peculiar problem which so far is only partially understood and which is discussed in Chapter 9. The author has recently made some progress on this problem, which will be treated in a future publication.

The style of this book is informal. Some mathematical texts start with a fivefold definition, followed by the statement of a theorem in a similar style, and in the end finally a proof. Usually no conclusion, further explanation or justification. In this book this sort of writing will be avoided.

More on the contents of this book

The first chapter discusses general features of the notion of an algebraic dynamical system, with a few remarks on the special cases in which, strictly speaking, a different term has to be used. The importance of probabilistic aspects is emphasized.

The second chapter is brief. It gives the main elements constituting an algebraic dynamical system. i.e. an algebra of observables, a state functional, the physical interpretation of both, time evolution and, finally, symmetries.

In the third chapter we illustrate the general concepts of the preceding chapter by describing in more detail how these are realized concretely, both in classical and in quantum physics.

Behind the ideas on the use of commutative as well of non-commutative topology and differential geometry, there is a fruitful general mathematical idea, which is called 'non-commutative geo-metry', a name due to its creator, the mathematician Alain Connes. This will be discussed in the fourth chapter.

Algebras, both commutative and non-commutative form an important ingredient of this book. The fifth chapter describes the properties of the various types of algebras. It is in fact a collection of 'mathematical intermezzos' on the notion of 'algebra', to be used in the other chapters

The sixth chapter is devoted to the GNS (Gelfand-Naimark-Segal) representation, important for the physical interpretation of pairs of observables and states.

The seventh chapter contains a discussion of quantum field theory, with its great success in elementary particle physics, and at the same time with its deep, still unsolved mathematical problems. The first serious approach to these problems, Wightman's formulation of quantum field theory is discussed.

Wightman theory has a natural formulation as an algebraic covariance system, a name proposed in Chapter 1. It is due to Armin Uhlmann and Hans-Jürgen Borchers. For this C^*- or von Neumann algebras are no longer sufficient. One needs more general locally convex topological algebras, so far very rarely employed in mathematical physics. This is explained in the eighth chapter.

We show in the nineth chapter that the special case of the Maxwell quantum field can be adequately described by a small but interesting generalization of the notion of algebraic covariance system in the Uhlmann-Borchers framework. The use of these more general algebras becomes crucial.

In the tenth chapter we discuss the more general notion of an algebraic quantum field theory, based on a system of C^*-algebras attached to finite open sets in spacetime. None of the authors on algebraic quantum field theory treat the case of so-called gauge theories, of which the Maxwell field is the simplest example. The reason for that is that because of the incompatibility between manifest Poincaré covariance and a state space with a positive definite inner product, such fields clearly do not fit in the standard form of algebraic quantum field theory.

A sketch of a generalized algebraic quantum field theory will be presented in which this problem can (hopefully) be solved. In this approach the theory consists of two parts, an 'unphysical' part, which involves spaces with an indefinite inner product, and a second 'physical' one, obtained by a quotient construction from the first. Both parts are necessary; the unphysical part is needed to formulate the dynamics. The theory uses also more general algebras. All this will be explained in Chapter 10.

Ordinary probability theory can be formulated in terms of commutative algebras. The non-commutative analogue is called quantum probability; it is in fact just quantum mechanics, seen from a somewhat unusual perspective.

In Chapter 11 we discuss briefly 'stochastic graph theory' and explore the non-commutative version of it. The basic idea of the commutative version is simple. Its sample space comes from a probabilistic model in physics, statistical mechanics, the phase of classical mechanics. It is defined as a collection of finite graphs of a certain type. One defines probability measures on this space in analogy with the classical ensembles. The results have a wide range of interesting applications outside physics, such as the quantum Black-Scholes equation in financial mathematics, proposed by Segal & Segal. We shall give a brief discussion of this rather unexpected example.

Remarks on references

The first chapter discusses general features of the notion of an algebraic dynamical system, with a few remarks on the special cases in which, strictly speaking, a different term has to be used. The importance of probabilistic aspects is emphasized.

The second chapter is brief. It gives the main elements constituting an algebraic dynamical system. i.e. an algebra of observables, a state functional, the physical interpretation of both, time evolution and, finally, symmetries.

In the third chapter we illustrate the general concepts of the preceding chapter by describing in more detail how these are realized concretely, both in classical and in quantum physics.

Behind the ideas on the use of commutative as well of non-commutative topology and differential geometry, there is a fruitful general mathematical idea, which is called 'non-commutative geo-metry', a name due to its creator, the mathematician Alain Connes. This will be discussed in the fourth chapter.

Algebras, both commutative and non-commutative form an important ingredient of this book. The fifth chapter describes the properties of the various types of algebras. It is in fact a collection of 'mathematical intermezzos' on the notion of 'algebra', to be used in the other chapters

The sixth chapter is devoted to the GNS (Gelfand-Naimark-Segal) representation, important for the physical interpretation of pairs of observables and states.

The seventh chapter contains a discussion of quantum field theory, with its great success in elementary particle physics, and at the same time with its deep, still unsolved mathematical problems. The first serious approach to these problems, Wightman's formulation of quantum field theory is discussed.

Wightman theory has a natural formulation as an algebraic covariance system, a name proposed in Chapter 1. It is due to Armin Uhlmann and Hans-Jürgen Borchers. For this C^*- or von Neumann algebras are no longer sufficient. One needs more general locally convex topological algebras, so far very rarely employed in mathematical physics. This is explained in the eighth chapter.

We show in the nineth chapter that the special case of the Maxwell quantum field can be adequately described by a small but interesting generalization of the notion of algebraic covariance system in the Uhlmann-Borchers framework. The use of these more general algebras becomes crucial.

In the tenth chapter we discuss the more general notion of an algebraic quantum field theory, based on a system of C^*-algebras attached to finite open sets in spacetime. None of the authors on algebraic quantum field theory treat the case of so-called gauge theories, of which the Maxwell field is the simplest example. The reason for that is that because of the incompatibility between manifest Poincaré covariance and a state space with a positive definite inner product, such fields clearly do not fit in the standard form of algebraic quantum field theory.

A sketch of a generalized algebraic quantum field theory will be presented in which this problem can (hopefully) be solved. In this approach the theory consists of two parts, an 'unphysical' part, which involves spaces with an indefinite inner product, and a second 'physical' one, obtained by a quotient construction from the first. Both parts are necessary; the unphysical part is needed to formulate the dynamics. The theory uses also more general algebras. All this will be explained in Chapter 10.

Ordinary probability theory can be formulated in terms of commutative algebras. The non-commutative analogue is called quantum probability; it is in fact just quantum mechanics, seen from a somewhat unusual perspective.

In Chapter 11 we discuss briefly 'stochastic graph theory' and explore the non-commutative version of it. The basic idea of the commutative version is simple. Its sample space comes from a probabilistic model in physics, statistical mechanics, the phase of classical mechanics. It is defined as a collection of finite graphs of a certain type. One defines probability measures on this space in analogy with the classical ensembles. The results have a wide range of interesting applications outside physics, such as the quantum Black-Scholes equation in financial mathematics, proposed by Segal & Segal. We shall give a brief discussion of this rather unexpected example.

Remarks on references

We noted already earlier in this Preface that this book has a system of *Intermezzos* for the knowledge of mathematical or physics that students may need. Extensive references can moreover be found in the reference sections of the various chapters. These are of three types. First we give what might be called the basic ones, usually books that can be considered to be standard references. They are often not of the most accessible sort, so the second type consists of introductory and more friendly books. Finally there are equally introductory lecture notes available on the internet. The internet addresses for these are provided. They may be ephemeral, disappearing after some time, but the presence of such lecture notes on the internet is in general remarkably stable.

Wikipedia articles, available on almost any topic, are useful and can in general be trusted. Useful for mathematical topics is the Encyclopedia of Mathematics, with a starting page which can be accessed at

`https://www.encyclopediaofmath.org/index.php/Main_Page`

Some topics in physics have philosophical aspects. For this the Stanford Encyclopedia of Philosophy can be recommended. Its starting page is at

`http://plato.stanford.edu/`

8

Its articles are authoritative and clearly written. It covers a wide range of topics. See, for example, the items "quantum mechanics" or "dynamical system".

For a general background to the mathematics of theoretical physics the magnificent book of Roger Penrose can be recommended [8].

A scientist in Kazachstan has a website, "Sci-Hub", with some 49 million scientific articles, hacked from various sources; so far 20 million have been downloaded, in particular in Iran, India and China, where scientists are unable to pay the excessive prices of legal downloads.

Final remarks

1. Physics and mathematics should not be taught along historical lines. Nevertheless a certain knowledge of the history of science is a matter of general education. For this reason historical details can be found throughout the text of this book. The history of physics is interesting, among other things because it illustrates that the development of science has never been straightforward, but was always full of wrong tracks, dead ends and correct results obtained by wrong arguments.

2. A physical theory is a mathematical model of a physical situation, always approximative. (It is in any case not completely clear what we mean by 'reality' or the 'real world', in particular since the discussions on the Einstein-Podolsky-Rosen paradox.) This model should be logically consistent. It should contain predictions that can be tested experimentally. If the result of such a test is negative, the theory is wrong; if it is positive then the theory is acceptable, for the time being, in the spirit of Popper's ideas on the philosophy of science. If the theory does not have predictions that can be tested experimentally – or observed, as in astronomy, then, allowing for a reasonable period of time, one has to admit that it is not a physical theory. If sufficiently imaginative it is science fiction. Examples of this are string theory in elementary particle physics and the many-world interpretation of quantum mechanics. For a critical discussion of these examples [9] and [10] are recommended.

3. Using a book for a course should not mean reading from it but using it as a background for further information and for the stimulation of the imagination of teacher and students.

4. The main merit of this book, if any, is that it puts known matters into a new framework, and does this in a manner which should be understandable for a general readership of physics and mathematics students. Remarks and corrections will be welcome at p.j.m.bongaarts[at]xs4all.nl.

5. Finally, a piece of information that is totally irrelevant to the subject matter of this book, but which the author, given his country of origin, finds amusing enough to impart to a wider international readership.

In all European languages, the names of mathematical concepts are – almost without exception – derived from Latin or Greek. Dutch is an exception. Most such names, except the very recent ones, are Dutch. Examples:

mathematics > *'wiskunde'* = knowledge which is sure,
geometry > *'meetkunde'* = art of measuring,
polynomial > *'veelterm'* = that which has many terms,
arithmetic > *'rekenkunde'* = art of calculation.

Etc. The inventor of this terminology was Simon Stevin (1548-1620), a Flemish-Dutch mathematician and engineer, who was a great admirer of Archimedes, invented practical machines and tools, and contributed in an important manner to the theory of mechanics. He also introduced Dutch terms for other subjects, e.g.

physics > *'natuurkunde'* = knowledge of physical phenomena,

geography > *'aardrijkskunde'* = knowledge of the earth.

He had the somewhat debatable conviction that Dutch was the language *par excellence* for doing science. For a short biography, see [11] (on internet).

References

1. I.E. Segal : Postulates for general quantum mechanics. Ann. of Math. 48, 930-948 (1947).
2. Rudolph Haag, Daniel Kastler : An algebraic approach to quantum field theory. J. Math. Phys. 5, 848-861 (1964).
3. Ludwig Faddeev, O.A. Yakubowskiĭ : Lectures on Quantum Mechanics for Mathematics Students. American Mathematical Society 2009.
4. Franco Strocchi : An Introduction to the Mathematical Structure of Quantum Mechanics. World Scientific 2005.
5. Peter Bongaarts : Quantum Theory. A Mathematical Approach. Springer 2014.
6. Peter Bongaarts : The Foundations of Quantum Theory. A Critical Assessment. In preparation.
7. T. Mol, W.F. den Hollander, D. Garlaschelli : Breaking of Ensemble Equivalence in Networks. Eurandom report 2015.
8. Roger Penrose : The Road to Reality. BCA 2004.
9. Peter Woit : Not Even Wrong: The Failure of String Theory and the Search for Unity in Physical Law. Basic Books 2007.
10. J. Baggott : Farewell to Reality: How Modern Physics Has Betrayed the Search for Scientific Truth. Pegasus 2014.
11. A. van Assche et al.: The story of Simon Stevin. No date. Available at
    `http://mathsforeurope.digibel.be/Stevin.htm`

ACKNOWLEDGEMENTS

I am much indebted to

- Henk Pijls, for checking the mathematics and for help with LaTeX.

- Vesselin Petkov, my friendly and competent editor.

- Terry Deveau, for corrections.

- My wife, for checking my English.

- The anonymous contributors of Wikipedia.

1 ALGEBRAIC DYNAMICAL SYSTEMS. INTRODUCTION

In this book an *algebraic dynamical system* is a physical system, classical or quantum, seen from the general viewpoint that the primary notion in describing such systems is the algebra of observables, commutative for classical systems, non-commutative for quantum systems, the secondary notion that of the state, with the third notion the physical interpretation of the combination of these two. There are two more basic notions, the fourth one for the dynamics, the time development, finally the fifth one which describes possible symmetries of the system.

The name 'algebraic dynamical system' is a general term; it also covers a few versions that deserve slightly different names because they do not have dynamics in the sense of time development.

According to the special theory of relativity space and time are not separate notions. The division of 4-dimensional spacetime into space and time depends on the chosen coordinate system and so does the idea of time development. What is intrinsically coordinate independent, is the covariance under inhomogeneous Lorentz transformations. This means that for this case one needs only symmetry as fourth basic idea. For this reason we shall describe relativistic physical systems in this context not as algebraic dynamical systems but as *algebraic covariance systems*. Typical examples of such systems will be discussed in Chapter 3 and later in Chapter 8 and Chapter 11.

Systems in equilibrium do not have a time development; there is no dynamics, but there may be symmetries. Such systems can be called *algebraic probability systems*. They will be treated in detail in later chapters.

Probabilistic aspects are of great importance in our formalism, even though for some cases these aspects are trivial. An example of this is classical mechanics. The algebra of observables is the algebra of all smooth functions on phase space, in general a symplectic manifold. States, as linear functionals on this algebra, are associated with δ-function-like point masses.

See for information on probability theory the mathematical intermezzo *Probability theory* - to be found in Chapter 3, Section 2, and for the δ-

function the mathematical intermezzo *Generalized functions, Dirac δ-function* - in Chapter 8, Section 3.

2 ALGEBRAIC DYNAMICAL SYSTEMS. MAIN ELEMENTS

2.1 Important notions

Our formalism has five basic notions: observable, state, the interpretation of these two, time evolution and symmetry. The first three are the most important.

2.2 Observables

The observables of a physical system are selfadjoint elements a, i.e. with $a^* = a$, of a complex $*$-algebra $\mathcal{A}$, associative and with unit element. Note that, strictly speaking, $\mathcal{A}$ should be called the *algebra of pre-observables*. The reason for this is that in some cases the true physical observables are the smallest von Neumann algebra of the representation of $\mathcal{A}$ in a physical Hilbert space, in particular the GNS representation space, which will be discussed in Chapter 6.

- Mathematical intermezzo

Vector spaces, linear algebra. An elementary notion, but a reminder of some properties may nevertheless be useful, in particular as introduction to the next intermezzo.

A vector space V is a set with two binary operations, addition of two elements, $V \to V$, $(x, y) \mapsto x + y$, and scalar multiplication $\mathbb{R} \times V \to V$ (for a real vector space), $\mathbb{C} \times V \to V$ (for a complex vector space), $(\lambda, x) \mapsto \lambda x$, with obvious additivity and associativity properties. From now on our vector spaces will be over the complex numbers, unless explicitly stated otherwise. For the complexification of a real vector space, see the mathematical intermezzo - Complexification of vector spaces and algebras to be found in Chapter 6, Section 2.

Important are linear transformations $T : V \to V$, or $T : V \to W$ when a second vector space W is involved. A vector space V has a dual vector space V^*, consisting of all linear functionals on V, i.e. linear maps $V \to \mathbb{C}$ (to $\mathbb{R}$ in the real case). The dual of the dual is denoted as $(V^*)^*$.

Problem. $V \subset (V^*)^*$. Prove this, and show that the equality sign holds for a finite dimensional V. Hint: use a dimension argument. Note that this property is called *reflexivity*. Some infinite-dimensional vector spaces are reflexive, for instance Hilbert spaces. See the mathematical intermezzo - Hilbert space in Chapter 3, Section 3.

The subject of vector spaces belong to *linear algebra*. References for this are [1], [2] and [3] (all three on internet).

- End of mathematical intermezzo

- Mathematical intermezzo - extra

Algebras. A real or complex vector space $\mathcal{A}$ provided with an additional binary operation, a multiplication $(a, b) \mapsto ab$, which is associative, i.e. $(ab)c = a(bc)$ – but see the exception to this below, and distributive with respect to the addition, i.e. $a(b + c) = ab + ac$, for all a, b, c in $\mathcal{A}$. An algebra can or may not have a unit element; but the algebras in this book all have a unit element.

For the complex case there is the notion of *-algebra, which means that there is a conjugate linear map $a \mapsto a^*$, with $(ab)^* = b^* a^*$, and of course $1_{\mathcal{A}}^* = 1_{\mathcal{A}}$. In principle all algebras in this book are complex, or can be complexified. See for this notion the mathematical intermezzo - Complexification of vector spaces and algebras to be found in Chapter 6, Section 2.

An algebra defined in this manner, i.e. as a set with certain operations is sometimes called an 'abstract algebra'. Algebras of $n \times n$ matrices are examples of 'concrete' algebras.

In Chapter 7 there will be an extensive discussion of the various types of algebras that play a role in physics, C^*-algebras, von Neumann algebras, Fréchet algebras, etc.. There are good general introductions to 'abstract' algebra in [4], [5] and [6] (all three on internet). The first gives also useful general information on set theory, vector spaces, matrices and groups.

An important notion in physical applications of algebras is that of *representations*. It is a realization of the elements of the algebra as linear transformations. Precise definition: a representation π of an algebra $\mathcal{A}$ in a vector space V is a linear map

$$\pi : \mathcal{A} \to GL(V), \quad a \mapsto \pi(a), \quad \forall a \in \mathcal{A},$$

with $GL(V)$ the algebra of all linear maps $V \to V$, and such that
  1. $\pi(ab) = \pi(a)\pi(b), \forall a, b \in \mathcal{A}$,
  2. $\pi(1_{\mathcal{A}}) = 1_V$.

For a *-algebra one has *-representations in a Hilbert space, (or pre-Hilbert space), i.e. linear maps

$$\pi : \mathcal{A} \to B(\mathcal{H}), \quad a \mapsto \pi(a), \quad \forall a \in \mathcal{A},$$

with $B(\mathcal{H})$ the algebra of all bounded operators in $\mathcal{H}$, and with the additional condition

3. $\pi(a^*) = (\pi(a))^*$,
4. $(\pi(a)\psi_1, \psi_2) = (\psi_1, \pi(a^*)\psi_2)$, $\forall a \in \mathcal{A}$, $\psi_1, \psi_2 \in \mathcal{H}$.

In physics one also employs the important example of Lie algebras, a non-associative algebra. See for this the mathematical intermezzo - Lie algebras to be found in this chapter, Section 6. In Mathematical intermezzo - extra - Operators in Hilbert space found in Chapter 3, Section 3, where the notion of bounded operator is explained.

In [7] (on internet) there is a detailed discussion of representations of algebras.

- End of mathematical intermezzo - extra

Basic idea that underlies this book: *Classical systems are described by commutative algebras; quantum systems by non-commutative algebras. Apart from that their structure is very similar.*

The two cases can be related by inserting Planck's constant $\hbar$ in $\mathcal{A}$. In one direction there is the *classical limit* of a quantum system, obtained by letting (the numerical value) $\hbar \to 0$. Indeed $\hbar$ is a constant of nature; it has a dimension, so its numerical value depends on the system of units that one has in mind. By changing these units, $\hbar$ can go to 0 or become very large. In the other direction one has deformation of the classical picture by means of $\hbar$ as deformation parameter.

Remark: The use of complex numbers is essential in quantum theory; classical physics is for the main part expressed in real numbers. A real algebra $\mathcal{A}$ can be complexified to a complex $*$-algebra. In turn a complex commutative $*$-algebra has a real subalgebra, consisting of the elements a with $a^* = a$. Both cases carry the same amount of information. The real part of a complex non-commutative $*$-algebra is *not* a subalgebra. So in our general formalism $\mathcal{A}$ will be a complex $*$-algebra. For a description of the construction of complexification of real vector spaces and algebras see the mathematical intermezzo - Complexification of vector spaces and algebras, to be found in Chapter 5, Section 2.

From now on all vector spaces and algebras will be assumed to be complex, unless the opposite is explicitly stated.

2.3 States

States of a system are expectation functionals, i.e. normalized positive linear functionals ω on $\mathcal{A}$, which means $\omega(1) = 1$, and $\omega(a^*a) \geq 0$, for all $a \in \mathcal{A}$.

18

2.4 Interpretation of 1 and 2

Choice of a state ω and an observable a leads to a Hilbert space $\mathcal{H}_\omega$, by means of the so-called *GNS representation* π_ω in a Hilbert space denoted as $\mathcal{H}_\omega$. (GNS = Gelfand-Naimark-Segal). Its construction and properties will be separately discussed in Chapter 7.

All this, and much of what will follow immediately, suggests quantum theory. However I shall show in the next chapter that classical physics also fits in this scheme, although in a somewhat less obvious and sometimes trivial way.

2.5 Time evolution

Time development of the system is described by $*$-preserving automorphisms $\{\phi_t\}_{t\in\mathbb{R}}$ of the algebra $\mathcal{A}$. A state ω with $\omega(\phi_t(a)) = \omega(a)$, for all $t \in \mathbb{R}$, is called an *invariant state*. An important consequence of this invariance is that the automorphisms ϕ_t are unitarily implementable in the GNS Hilbert space $\mathcal{H}_\omega$, emerging in the GNS construction, mentioned in Section 3, and to be discussed in detail in Chapter 7. This means that there exists a 1-parameter system of unitary operators $\{U_\omega(t)\}_{t\in\mathbb{R}}$, in $\mathcal{H}_\omega$ such that

$$\pi(\phi_t(a)) = U(t)\pi(a)(U(t))^{-1}, \quad \forall t \in \mathbb{R}, \ \forall a \in \mathcal{A}.$$

An observable a such that $\phi_t(a) = a$, for all $t \in \mathbb{R}$, is called a *constant of motion*.

We may assume that the 1-parameter group of automorphisms $\{\phi_t\}_{t\in\mathbb{R}}$ is continuous in t, in an appropriate sense. Then the operators $U(t)$ can be written as $U(t) = e^{itL}$, with the L the selfadjoint generator of the $U(t)$, in fact what in the quantum case will be the Hamiltonian operator, up to a sign and to a factor $\hbar$.

Remark: A physical system with the time development as described here, is sometimes called an *autonomous system*. A more general case is that of a *non-autonomous system*, in which the time evolution is given by a 2-parameter system $\{\phi_{t_1,t_2}\}_{t_1,t_2\in\mathbb{R}}$. Instead of the relations

$$\phi_0 = 1_\mathcal{A}, \quad \phi_{t_1+t_2} = \phi_{t_1}\phi_{t_2}, \quad \forall t_1, t_2 \in \mathbb{R},$$

one has

$$\phi_{t,t} = 1_\mathcal{A}, \quad \phi_{t_3,t_2}\,\phi_{t_2,t_1} = \phi_{t_3,t_1}, \quad \forall t_3, t_2, t_1 \in \mathbb{R}.$$

One says that the interaction in such a situation is time-dependent. This case will not be discussed further in this book.

Remark. As we have already noted in Chapter 1, time development is not an intrinsic property in systems in special relativity, so it is therefore absent. See for special relativity [8] and [9] (both on internet). In this case symmetry remains. We called such systems 'algebraic covariance systems'. For these there is in the first place symmetry with respect to the inhomogeneous Lorentz group, also called the Poincaré group. See the mathematical intermezzo - Poincaré group or inhomogeneous Lorentz group, to be found in the next section. Additional symmetries may occur. We also discussed systems in equilibrium, i.e. algebraic dynamical systems which do not evolve in time. We called these 'algebraic probability systems'. Again time development is missing. There may be symmetries, but there is no basic symmetry like in relativistic systems.

2.6 Symmetries

Symmetries are very important in physics, in quantum physics even more than in classical physics. A symmetry of an algebraic dynamical system is described by a group of $*$-automorphisms $\phi(g)_{g \in \mathcal{G}}$ with $\mathcal{G}$ the symmetry group of the system. These $*$-automorphisms should commute with the time development $*$-automorphisms ϕ_t and leave the state ω invariant.

As a consequence these automorphisms are also unitarily implementable in the GNS-Hilbert space $\mathcal{H}_\omega$.

Problem. Use the analogy with the situation in Section 4 to say what this means.

- Mathematical intermezzo

Groups. A group $\mathcal{G}$ is a set with a multiplication between elements, which is associative, i.e, with $(g_1 g_2)g_3 = g_1(g_2 g_3)$. $\mathcal{G}$ has a unit element; each element has an inverse.

This definition describes an 'abstract' group, i.e. a set with a certain binary operation, just as in the earlier case of algebras. A group of $n \times n$ matrices is an example of a 'concrete group'. Most groups in physics are indeed matrix groups.

A group, like an algebra, can be *represented* by linear transformations in a vector space. This is a very important notion in the application of group theory in physics. It means that for every element g of a group $\mathcal{G}$ one has a linear transformation $\pi(g)$ in a vector space V such that $\pi(g_1 g_2) = \pi(g_1)\pi(g_2)$ and $\pi(1_\mathcal{G}) = 1_V$. Of particular importance are *unitary representations* by unitary operators in a Hilbert space. The representation theory of groups is a nontrivial appendix to group theory.

There are finite groups, i.e. with a finite number of elements; the usual groups in physics are infinite, i.e. continuous or Lie groups, with the underlying structure of a differential manifold. See for the notion of differentiable manifold the mathematical intermezzo - Differentiable manifolds, differential geometry, to be found in Chapter 5, Section 4.

Among the many good introductory books on group theory we may mention the books by Hall [10] and by Rothman [11]. A good set of lecture notes

20

available on the internet is [12]. References for representation theory are [13] and [14] (both on internet).

- End of mathematical intermezzo

- Begin of mathematical intermezzo

Lie algebras. Vector spaces with an additional binary operation, called a bracket, $(x, y) \mapsto [x, y]$, which is antisymmetric, bilinear in the two variables, and satisfies the Jacobi identity

$$[x, [y, z]] + [y, [z, x]] + [z, [x, y]] = 0.$$

Defined in this manner it is an 'abstract' Lie algebra; a matrix Lie algebra, consisting of $n \times n$ matrices is a concrete Lie algebra. In this case the bracket is just the commutator $[x, y] = xy - yx$.

The theory of *representation of Lie algebras* is important in its application to quantum theory, even more so than in the case of group representations, because of the linearity of Lie algebras.

Good lecture notes on Lie algebra theory are [15] and [16] (both on internet).

Problem. Show that the collection of all linear transformation in a vector space is a ('concrete') Lie algebra. There is an intimate connection between Lie groups and Lie algebras, which is one-to-one if one passes over some nontrivial mathematical details. This is particularly easy to understand for matrix groups. A matrix L from the Lie algebra gives a 1-parameter subgroup $G(t)$ of the Lie group, by exponentiating according to $G(t) = e^{tL}$.

- End of mathematical intermezzo

Group theory has shown itself to be very useful in understanding quantum mechanics and in particular atomic physics. The main group there is $SO(3)$, the rotation group in 3-dimensional space. For more information on this group and its representations, see [17] and [18] (both on internet).

The history of the mathematical understanding of quantum mechanics starts with John von Neumann, whose classic book "Mathematical Foundations of Quantum Mechanics" [19] laid the basis for the Hilbert space framework of the theory. Second came group theory, with fundamental work by Bartel Leendert van der Waerden and in particular Hermann Weyl. Weyl's book is another historical classic [20] (complete on internet).

The mathematical contributions to quantum mechanics, as those of Van der Waerden and Weyl, were not always welcome in the physics community; they were often considered to be superfluous and too abstract. Physicists, in particular experimental spectroscopists, spoke of "Gruppenpest"("Group Plague").

- Mathematical intermezzo

Poincaré group or inhomogeneous Lorentz group. One writes the elements of the Poincaré group as pairs (a, Λ), with $a \in \mathbb{R}^4$ a translation in 4-dimensional spacetime, and Λ a 4×4 matrix. These matrices, which are supposed to act on vectors $x = (x^0, x^1, x^2, x^3)$ of spacetime, leaving the indefinite quadratic form $(x^0)^2 - (x^1)^2 - (x^2)^2 - (x^3)^2$ invariant, form the homogeneous Lorentz group, which is often denoted as $O(1,3)$. An element (a, Λ) of the Poincaré group acts on spacetime as

$$((a, \Lambda)x)^\mu = \sum_{\nu=0,1,2,3} (\Lambda^\mu_\nu x^\nu + a^\nu),$$

or in a shortened notation

$$(a, \Lambda)x = \Lambda x + a.$$

Problem. Show that the product of two group elements is equal to

$$(a_2, \Lambda_2)(a_1, \Lambda_1) = (a_2 + \Lambda_2 a_1, \Lambda_2 \Lambda_1),$$

and that

$$(a, \Lambda) = (-\Lambda^{-1} a, \Lambda^{-1}).$$

The Poincaré group is the hallmark of special relativity. For a short overview of the properties of the Lorentz and Poincaré groups, see [21].

For remarks on relativistic conventions and notation, see the mathematical intermezzo *Relativistic conventions and notation* - to be found in Chapter 9, Section 1.

- *End of mathematical intermezzo*

2.7 Algebraic dynamical systems. A slight generalization

In our standard notion of an algebraic dynamical system we consider pairs of classical systems with commutative algebras of observables and quantum systems with non-commutative algebras of observables.

This applies to quantum theory as developed in the time of Heisenberg and Schrödinger. However in 1924 *spin*, an intrinsic angular momentum of in particular the electron was discovered by Wolgang Pauli, who did not publish his observation, and in 1925 by George Uhlenbeck and Samuel Goudsmit, who did. Since then we know that there are two kind of particles, which cannot be transformed into each other, *bosons*, particles with integer spin, and *fermions*, with half-integer spin. The electron has spin $1/2$. With boson quantum systems there are corresponding classical theories with commutative algebras of observables. There are no physically meaningful classical fermion systems, but only unphysical 'pseudo-classical' ones.

Nevertheless pseudo-classical systems play an important role as auxiliary notions. For calculations of scattering amplitudes in elementary particle physics the very effective and also very heuristic method of path integration is used, in which (infinite-dimensional) integrals over spaces of classical fields appear. This is straightforward for boson fields. For fermion fields there is the so-called Berezin integral over 'anti-commuting c-numbers 'or 'Grassmann variables'.

- Mathematical intermezzo

Grassmann variables, Berezin integral - Let V be a vector space and define as the Grassmann algebra over V

$$T_a(V) = \oplus_{n=0}^{\infty} (\otimes^n V)_a$$

which is of course the antisymmetric tensor algebra over V. For the notion of tensor product of vector spaces, see the Mathematical intermezzo *Tensor product. Tensor algebra* - to be found in Chapter 8, Section 2. Each element a in $T_a(V)$ can be uniquely written as a direct sum

$$a = a^0 + a^1,$$

in which a^0 is called the *even part* of a, and a^1 the *odd part*. It is not hard to check the following multiplication rule

$$a^j b^k = (-1)^{j+k} b^k a^j, \quad \forall a, b \in T_a(V).$$

$T_a(V)$ is built from single elements a. These are sometimes called *anti-commuting c-numbers* or *Grassmann numbers*. They behave like ordinary complex numbers, except for what is usually called their 'graded commutativity or 'almost commutativity'. See [22] (on internet). They play an important part in the pseudo-classical version of fermion systems.

An important application is *Berezin integration*, a linear functional on the algebra $T_a(V)$. An infinite dimensional version is used in particle physics as a path integral over pseudo-classical fermion fields.

See for the notion of Grassmann algebra [23], Grassmann analysis [24], for anti-commuting variables [25] and Berezin integration [26] (the last three on internet).

- End of mathematical intermezzo

Grassmann variables appear in a whole new field of mathematics, with topics all having the prefix 'super': super-differential geometry, super-manifolds, super-Lie algebras, all playing a role in physical theories like supersymmetry, super-gravity, super-strings. Although the effects predicted by these theories have so far not been detected experimentally, the mathematics remains of interest.

References

1. Cherney, D, Denton, T., Waldron, A.: Linear Algebra. Lectures at the University of California at Davis 2013. Accessible at

https://www.math.ucdavis.edu/~linear/linear-guest.pdf
2. Dawkins, P. Linear Algebra. Lectures at Cornell University 2005. Accessible at
http://www.cs.cornell.edu/courses/cs485/
2006sp/linalg_complete.pdf
3. Hefferson, J. : Linear Algebra. Lectures. No date. Accessible at
http://joshua.smcvt.edu/linearalgebra/book.pdf
4. Malik, D.S, Mordeson J.N., Sen, M.K.: Introduction to Abstract Algebra. Lectures at Creighton University 2007. Avalaible at
https://people.creighton.edu/~dsm33733/
MTH581/Introduction%20to%20Abstract%20Algebra.pdf
5. Walker, E.A.: Introduction to Abstract Algebra. Random House 1987. Revised as a series of lectures at the State University of New Mexico 1998. As such available at
https://www.math.nmsu.edu/~elbert/AbsAlgeb.pdf
6. Joyce, D.E.: Introduction to Modern Algebra. Clark University lecture notes 2008. Available at
http://aleph0.clarku.edu/~djoyce/ma225/algebra.pdf
7. Wallach, N.R.: Algebras and Representations. University of California at San Diego lectures. No date. Available at
http://www.math.ucsd.edu/~nwallach/chapter4.pdf
8. Cresser, J.D.: Lecture Notes on Special Relativity. Macquarie University 2005. Available at
http://physics.mq.edu.au/~jcresser/Phys378/
LectureNotes/VectorsTensorsSR.pdf
9. Komissarov, S.S.: Special Relativity. Leeds University Lecture Notes 2012. Available at
https://www1.maths.leeds.ac.uk/~serguei/
teaching/gr.pdf
10. Hall, B.: Lie groups, Lie Algebras, and Representations: An Elementary Introduction. Springer 2004.
11. Rothman, J.: An Introduction to the Theory of Groups. Springer 1999.
12. Milne, J., D.: Group Theory. 2012. Available at:
http://www.jmilne.org/math/CourseNotes/GT.pdf
13. Vvedensky, D : Representations of Groups. Imperial College lecture notes. No date. Available at
http://www.cmth.ph.ic.ac.uk/people/
d.vvedensky/groups/Chapter3.pdf
This is Chapter 3 of a full course on group theory, avalaible at
http://www.cmth.ph.ic.ac.uk/people/d.vvedensky/groups/
14. Smith, K: Groups and their Representations. University of Michigan lecture notes. No date. Available at
http://www.math.lsa.umich.edu/~kesmith/rep.pdf
15. NN: Lie Algebra. Trinity College Dublin lecture notes. No date. Available at

```
http://www.phys.nthu.edu.tw/~class/
Group_theory/Chap%209.pdf
```
16. Samelson, H.: Notes on Lie Algebras. Springer 1990. As complete book available at
```
https://www.math.cornell.edu/~hatcher/Other/
Samelson-LieAlg.pdf
```
17. Avramidi, I: Notes on Groups SO(3) and SU(2). New Mexico Institute of Mining and Technology Lecture Notes 2011. Available at
```
http://infohost.nmt.edu/~iavramid/notes/su2.pdf
```
18. Soper, D.E.: The rotation group and quantum mechanics. University of Oregon lecture notes 2012. Available at
```
http://pages.uoregon.edu/soper/
QuantumMechanics/spin.pdf
```
19. von Neumann, J.: The Mathematical Foundations of Quantum Mechanics. Translated from the German. Princeton 1955.
20. Weyl, H.: The Theory of Groups and Quantum Mechanics. Translated from the 1930 German edition. Dover 2003. The complete book is available at
```
http://www.fulviofrisone.com/attachments/article/
485/Weyl,%20Robertson%20-%20The%20Theory%20O
f%20Groups%20&%20Quantum%20Mechanics.pdf
```
21. Yeh, N.: The Lorentz group and the Poincaré group. No date.
22. Bongaarts, P.J.M., Pijls, H.G.J. : Almost commutative algebra and differential calculus on the quantum hyperplane. J. Math. Phys. 35, 959-970 (1994). Available at
```
https://pure.uva.nl/ws/files/715978/186_2465y.pdf
```
23. Schulz, W.C. : Theory and application of Grassmann Algebra. University of Edinburgh Lecture notes. 2001. Accessible at
```
https://www.cefns.nau.edu/~schulz/grassmann.pdf
```
24. NN : Grassmann analysis: basics. Arizona State University. No date. Accessible at
```
http://swc.math.arizona.edu/dls/DLSCartierCh9.pdf
```
25. NN : Anticommuting variables. Edinburgh University lecture notes. Accessible at
```
http://www2.ph.ed.ac.uk/~rhorsley/SII10-11_mqft/lec09.pdf
```
26. Novak, S. : Berezin Integration. University of Hamburg lecture notes. 2011. Accessible at
```
http://www.math.uni-hamburg.de/home/sachse/berezin.pdf
```

3 MAIN ELEMENTS. EXPLICIT FORM

3.1 Introduction

In this chapter we explain in some detail the explicit form of the main elements of the scheme for algebraic dynamical systems introduced in the preceding chapter, both for the classical case and the quantum case. A series of topics from physics will be briefly reviewed, classical physics in general, classical statistical mechanics, thermodynamics, quantum mechanics. Added to these are mathematical intermezzos on and integration, probability theory, Hilbert space, operators in Hilbert space and the spectral theorem. Information on these subjects is also provided through the internet links given in the reference section.

3.2 Classical physics

- Physics intermezzo.

Classical physics. Because this book presents and discusses a general formalism for the uniform description of classical and quantum physics, two at first sight very different physical theories, it makes sense to give a brief sketch of what we mean by classical physics. This is the purpose of this intermezzo.

By classical physics one means the totality of physical theories that towards the end of the nineteenth century were supposed to describe all known phenomena of the physical world. Classical physics was based on two pillars, classical mechanics and classical electromagnetism, together with a description of the interaction between the two.

Classical mechanics gives a description of the forces between particles and the motion that is caused by these forces. The foundations were laid by Isaac Newton in the seventeenth century, with somewhat later further contributions by Joseph-Louis Lagrange and Pierre-Simon de Laplace, by William Rowan Hamilton in the nineteenth century, and finally by Henri Poincaré in the beginning of the twentieth century. The basic equations are a set of ordinary differential equations, depending on the particular point of view, Newton's, Lagrange's or Hamilton's equations, all three equivalent.

A state of a typical classical mechanical system, a system of N non-relativistic particles, is described by a point on the $6N$-dimensional phase space of the system, the space of $3N$ position and $3N$ momentum variables. The algebra of observables consists of the functions of these variables, which form a *commutative* algebra under pointwise multiplication. The positive linear functional ω from our general scheme is a point measure in a point $(\mathbf{p}_1, \ldots, \mathbf{p}_N, \mathbf{q}_1, \ldots, \mathbf{q}_N)$. Applying this functional on the algebra of observables gives, of course, this point. In this way such a system fits in our general algebraic framework, even though describing this case in this manner is overdoing it a bit. For the statistical description of a system of N classical particles, in particular if N is very large, the framework of Algebraic Dynamical Systems, i.e. in *classical statistical mechanics*, makes more sense, as will be clear from the physical intermezzo further in this section.

Good books on classical mechanics are those by Goldstein [1] (complete on internet), Kibble [2] and Taylor [3] (complete on internet). Useful introductory lecture notes are [4] (on internet).

The theory of electromagnetism describes electric charges, their attraction and repulsion, electric currents, magnetism, etc. James Clerk Maxwell collected all this into a single uniform theoretical framework, his general theory of electromagnetism, which is described by a set of partial differential equations.

For a short introduction to Maxwell's theory of electromagnetism see [5], for a more comprehensive set of lecture notes [6] and for a beautiful paper on the importance of Maxwell's work by Freeman Dyson, in which he explains why it took some twenty years before the physics community understood and appreciated Maxwell's work, see [7] (all three items on internet).

Important contributions to the understanding of the interaction between classical mechanics and electromagnetism were made by Hendrik Antoon Lorentz.

Towards the end of the nineteenth century, two seemingly small but persistent problems appeared. The first problem was the discreteness of atomic spectra, the second a problem with the propagation of light. Classical physics could not cope with these problems. The resolution of these problems came from the emergence of two new theories, quantum mechanics which explained the discrete atomic spectra, and the special theory of relativity which solved the problem of light propagation. The first is necessary for subatomic physics, the second for motion at very high velocities. In the macroscopic world, at 'ordinary' velocities, classical physics remains an extremely good approximation.

- End of physics intermezzo

- Physics intermezzo.

Classical statistical mechanics. If the number N for a classical system of N particles becomes very large, in the order of 10^{23}, as in a container filled with a gas, calculating the time development of the system becomes technically impossible, and would moreover be without interest. Instead, one calculates average quantities.

This means it becomes (classical) *statistical mechanics*, a subject founded by J. Willard Gibbs (1839-1903), the first internationally known American theoretical physicist, with contributions from Maxwell and Ludwig Boltzmann. Gibbs's book "Elementary principles in statistical mechanics" [8] (complete on internet) is a classic.

The mathematical basis of classical statistical mechanics is probability theory, which in turn is a special case of the theory of measure and integration. (See for these two subjects the mathematical intermezzos further down in this section), This basis is to be found in the first place in the work of Alexandr Yakovlevich Khinchin [9]. He was in fact very critical of the general level of mathematical rigour of theoretical physics, and in particular of statistical mechanics. Note in this connection the following quote from the preface of his book: "In the books on physics the formulation of the fundamental notions of the theory of probability as a rule is several decades behind the present scientific level."

The central notion of classical statistical mechanics is that of an 'ensemble". which is nothing but a probability distribution on the phase space of the $6N$ position and momentum variables of the classical N-particle system.

Various ensembles are used in classical statistical mechanics; they all give the same average values for physical quantities, in the so-called thermodynamic limit, i.e. in the limit for N going to infinity. The most important ensemble is the *canonical ensemble*, which describes a system contained in an infinite reservoir, a heat bath, which keeps it at a constant absolute temperature.

The final goal of statistical mechanics is to give, starting from the microscopic behaviour of a system of many particles, a rigorous derivation of *thermodynamics*, the description of physical systems in terms of temperature, pressure, volume, a phenomenological theory formulated long before the existence of atoms was generally accepted, with as its main applications to thermic devices such as steam engines. This goal has so far not fully been reached.

Good books on statistical mechanics are those by Huang [10] (complete on internet) and Van Vliet [11]; useful lecture notes are those of Gallavotti [12], Fitzpatrick [13], Frigg [14] and Huan [15] (all four on internet). The mathematical foundations are discussed by Dobrushin [16] (on internet). Fitzpatrick's notes contain a discussion of thermodynamics, the phenomenological theory of matter that statistical mechanics, the microscopical theory of matter, tries to explain. See also [17] (on internet).

It was already noted that an algebraic formulation for classical statistical mechanics is less farfetched than for classical mechanics. There is a one-to-one correspondence between probability theories and commutative von Neumann algebras provided with a positive linear (expectation) functional. Von Neumann algebras are discussed in the intermezzo - Von Neumann algebras, to be found in Chapter 6, Section 4.

- End of physics intermezzo

3.3 Mathematical information

- Mathematical intermezzo - extra.

Measure and integration. A *measurable space* is a pair $(X, \mathcal{B})$, with X a nonempty set, $\mathcal{B}$ the *measurable sets*, i.e. a system of subsets of X, satisfying the following convenient system of axioms:

(1). Both X and the empty set $\emptyset$ belong to $\mathcal{B}$.

(2). If A belongs to $\mathcal{B}$ than also its complement A^c.

(3). If the elements of the sequence $A_1, A_2, \ldots$ belong to $\mathcal{B}$, then so does their union.

The next step is a *measure space*, a triple $(X, \mathcal{B}, \mu)$, with a *measure* μ, i.e. a map from $\mathcal{B}$ to the nonnegative real numbers, which may include positive infinity.

The essentially bounded measurable functions form a commutative algebra by pointwise addition and multiplication. Essentially bounded means bounded, except at a set of measure zero. Note that for a property to hold 'except for a set of measure zero' means that the objects in question are the same, in the sense of measure theory. This occurs frequently. One usually says in this case that a property "holds for almost every point", or "almost everywhere".

A map f from one measure space $(X_1, \mathcal{B}_1, \mu_1)$ to a second measure space $(X_2, \mathcal{B}_2, \mu_2)$ is called measurable if and only if the inverse image $f^{-1}(A_2)$ of every element A_2 from $\mathcal{B}_2$ is an element of $\mathcal{B}_1$.

We may also consider complex-valued measurable functions, i.e. with the real and imaginary part both measurable.

There is an algebraic formulation of measure theory, in which an integral is a linear functional on the (commutative) algebra of measurable functions. A special case of this is a so called Radon measure. See [18], [19] and [20] (all three on internet). A classic on measure theory is the book of Halmos [21]; introductory courses are [22] and [23] (both on internet).

We shall see below that measure theory provides the basis for *Lebesgue integration*, a more advanced sort of integration than Riemann integration, used in elementary textbooks. Physicists are usually not familiar with the Lebesgue integral.

For the sake of simplicity we restrict the discussion here to nonnegative real functions on a finite interval of the real line. The higher dimensional case is a simple generalization of this and is therefore left to the reader. The general idea of integrating a function – in the 1-dimensional case – is the calculation of the surface beneath the function. Arbitrary functions can be written as a sum of a nonnegative and a negative function. The integral is then obviously the sum of the two separate integrals. Integrating a complex-valued function is just separately integrating the real and imaginary parts and adding the results.

The Riemann integral is an elementary part of standard analysis; it can stand on its own and needs no further mathematical background, unlike the Lebesgue integral, for which one has to start with the far from elementary theory of measure. But Lebesgue integration can do more, not only integrating over (basically) finite parts from $\mathbb{R}^n$ like the Riemann integral, but much more generally over arbitrary measure spaces. Riemann integration makes sense for bounded continuous (or piecewise continuous) functions. The Lebesgue integral can handle the much wider class of measurable functions. Also, improper integrals are simply included in the formalism.

The Riemann integral has a an intuitively rather obvious definition. One divides the (finite) interval of the real line over which one is integrating the function into a large number of subintervals. For such a partition there is an *upper sum*, the sum of the surfaces of all the rectangles above the subintervals, with height the supremum of the function above each subinterval. There is a similar notion of *lower sum*. Note that the *supremum* of a set of numbers is defined as the smallest number larger or equal to all numbers in the set. Equally the *infimum* is the largest number smaller or equal than all the numbers of the set. Obviously, for every partition the lower sum is smaller than or equal to the upper sum. If one lets the number of subintervals go to infinity, upper and lower sum may go to a limit. If moreover the two limits are the same, one says that the function is Riemann-integrable, with the common limit the Riemann integral of the function. Improper integrals, integrals over an infinite part of the real line, are defined as limits, i.e.

$$\int_0^{+\infty} f(x)dx = \lim_{a \to +\infty} \int_0^a f(x)dx,$$

provided that this limit exists. A similar limit procedure is used for the integration of unbounded functions. For a good introduction to the Riemann integral, see [24] (on internet).

The Lebesgue integral is rather different. It is defined for measurable functions on a given measure space $(X, \mathcal{B}, \mu)$. We first restrict the discussion to nonnegative (real-valued) functions on X and then consider a convenient class of functions for which the Lebesgue integral has an obvious meaning. Given a set A from $\mathcal{B}$, then the *indicator function* f_A is the function which takes the value 1 on A and is 0 elsewhere. A *simple function* is a linear combination of a finite number of indicator functions. Simple functions are obviously measurable. Let $(A_1, A_1, \ldots, A_n)$ be a sequence of sets from $\mathcal{B}$, and f_{A_j} the corresponding indicator functions, with the simple function $f(x) = \sum_j a_j f_{A_j}(x)$, and the a_j nonnegative numbers. Then the Lebesgue integral of f is defined as $\int_X f(x) d\mu = \sum_j a_j \, \mu(A_j)$. The simple functions form a normed vector space with norm $||f|| = \sum_j a_j \, \mu(A_j)$. For the notion of normed vector space, see the mathematical intermezzo *Hilbert Space*, Section 3.3 of this chapter. This space can be completed; the elements of this completed space are by definition Lebesgue integrable, with integral $\int_X f(x) d\mu = \sum_{j=1}^{\infty} a_j$. Note that this is a nonnegative number and may be infinite. One next does the same for nonpositive real-valued functions.

Combining these two results, one gets three possibilities:

• Both terms are finite. Then the integral is finite.

• One of the terms is finite; the other infinite. The integral is plus or minus infinity.

• Both terms are infinite. The integral is not defined.

Good lecture notes on Lebesgue integration, which, as in this text, explain the connections but also the differences with Riemann integration, are [25] and [26] (both on internet).

There are two related notions, Riemann-Stieltjes and Lebesgue-Stieltjes integral. In this situation two functions are involved, the function that is to be integrated, and a second function that is called an integrator. In the case of a Riemann-Stieltjes integral for a real-valued function of a single variable this integral has the form $\int_{\mathbb{R}} f(x) dg(x)$. For g differentiable, this is just $\int_{\mathbb{R}} f(x) \frac{d}{dx} g(x) dx$. However, this integral makes sense for more general functions, for example monotonically non-decreasing functions. For the Lebesgue-Stieltjes integral one has something similar.

See for useful lecture notes on Riemann-Stieltjes integration [27], and for Lebesgue-Stieltjes integration [28] (both on internet).

An operator-valued form of the Riemann-Stieltjes integral is used in the formulation of the spectral theorem for selfadjoint operators, the theorem that is central to the mathematical formulation of quantum mechanics.
See for the spectral theorem the intermezzo *Spectral theorem* and for quantum mechanics the intermezzo *Quantum mechanics*, both in this chapter, Section 3.3.

- End of mathematical intermezzo - extra

Probability theory just a particular case of measure theory. It plays a central role in this book.

- Mathematical intermezzo.

Probability theory. The history of probability goes back to the seventeenth century, with, among others, Pierre de Fermat and Blaise Pascal. Its main applications then were gambling and insurance. The modern mathematical formulation is due to Andrey Kolmogorov, formulated in the thirties of the last century [29] (complete on internet). A precursor was Gibbs, who laid the basis of statistical mechanics in his 1902 book, as was already mentioned.

It should be noted that some of the terminology introduced by Gibbs long before the work of Kolmogorov still persists in the physics literature. As a physics student, attending a course on statistical mechanics, it took the author a long time to realize that an 'ensemble' is just a probability distribution on phase space.

The books of Kolmogorov and Gibbs are classics and still worth reading.

A measure space $(X, \mathcal{B}, \rho)$ with $\rho(X) = 1$ is a probability space. Although the mathematical structure is the same, the terminology is different. The set X becomes Ω, measurable sets are called *events*, the set of all events is denoted as $\mathcal{F}$, the measure ρ becomes the probability P, measurable functions become *random variables*; a probability theory is then a triple $(\Omega, \mathcal{F}, P)$.

The simplest case is discrete probability. The event space Ω is an at most countable set of discrete points $\{x_n\}_n$, with probabilities P_n, with of course $\sum_n P_n = 1$. The average, mean or expected value for the random variable α is $E(\alpha) = \overline{\alpha} = \sum_n \alpha_n P_n$.

The next case is probability with a probability density function, which means, for the case where $\Omega = \mathbb{R}^n$, that there is a non-negative function $\rho(\mathbf{x})$ on $\mathbb{R}^n$ with

$$\int_{\mathbb{R}n} \rho(\mathbf{x})d\mathbf{x} = 1.$$

The expected value of the random variable α is $E(\alpha) = \int_{\mathbb{R}^n} \alpha(x)\rho(\alpha)dx$, etc. Finally there are so-called non-continuous probability theories which do not have a distribution function. We do not discuss these here.

The standard deviation, an important quantity describing the spread of the probability distribution, is $\sigma = \sqrt{E[(\alpha - E(\alpha))^2]}$.

A good introductory book on probability is Chung [30]; more comprehensive is Klemke [31]; Dudley [32] is advanced but very clearly written.

Useful lecture notes are [33], [34], [35] and [36] (all on internet).

There is a fairly obvious algebraic version of probability theory. Given an ordinary probabilistic system $(\Omega, \mathcal{F}, P)$ one defines a corresponding algebraic system $(\mathcal{A}, \omega)$, with $\mathcal{A}$ the commutative von Neumann algebra consisting of all essentially bounded, i.e. almost everywhere bounded, measurable functions on Ω and ω a positive normed linear functional on $\mathcal{A}$. For an element a in $\mathcal{A}$, the expression $\omega(a)$ is the expectation of a. The non-commutative analogue of this is immediately obtained by taking a non-commutative von Neumann algebra. This gives us the mathematical framework for quantum theory.

- End of mathematical intermezzo

3.4   Quantum physics 1

Quantum theory dominates the world of atomic and   subatomic physics. But it is also very much present in our daily life; without it there would be no television sets, mobile phones, computers and tablets. In our presentation here we shall distinguish in succession three stages, each a generalization of its predecessor. The first is quantum mechanics, describing systems of N non-relativistic particles, the second, quantum statistical mechanics for such a system when N is very large, the third relativistic quantum field theory, used in elementary particle physics. We start with a short review of quantum mechanics.

- Physics intermezzo - extra.

Quantum mechanics. New theories solve problems within existing theoretical frameworks. Quantum mechanics solved the problem of the discrete atomic spectra.

The prologue to quantum mechanics was the result of the experimental work of Ernest Rutherford in Manchester. His – in our modern eyes – simple small-scale scattering experiments led in 1911 to the picture of an atom as a small planetary system with negatively charged electrons encircling a positively charged nucleus. The hydrogen atom has a single electron; other atoms have more electrons.

Circular motion is accelerated motion. According to the insights due to Maxwell and Lorentz in this motion the electron emits radiation, looses kinetic energy and finally falls onto the nucleus. This would mean the emission of a short flash of light, not a constant discrete spectrum.

The young Niels Bohr, working with Rutherford in Manchester, taking his experimental results as a point of departure, postulated that the electron indeed moves in fixed orbits, but once in a while jumps to a lower orbit, emitting a precise amount of radiation corresponding with the energy difference of the two orbits. Bohr had no theoretical justification or background for his postulate; but it turned out to be a golden idea; he was able to calculate on the basis of it the precise spectrum of hydrogen.

During a few years his idea was further successfully applied to other atoms. This is what is sometimes called the 'old quantum theory'.

Finally, between 1923 and 1927 a complete new theory emerged, developed, by, among others, Werner   Heisenberg   and   Erwin Schrödinger. Like classical mechanics, it could describe the motion of a system of non-relativistic particles, but did this in a totally new and different manner. Bohr's postulate followed from it as a natural consequence. It may be called 'quantum mechanics proper'.

Originally, quantum mechanics came in two varieties. Heisenberg's matrix mechanics, using non-commuting observables, later identifies as infinite matrices. Schrödinger's wave mechanics, with as basic object his wave function, which describes the state of a quantum system. Soon it was shown that both theories were equivalent; different mathematical representations of the same theory.

The properties of quantum mechanics are very different from those of classical mechanics.

Quantum mechanics is *non-deterministic*. If in classical mechanics position and momentum of a particle are known at an initial time, then its position and momentum at any later time can – in principle – be calculated. Quantum mechanics can only predict the probabilities for the outcome of measurements.

Quantum mechanics is in its essence a probabilistic theory.

Albert Einstein, who made important contributions in the early phase of the theory, never accepted this. To him quantum mechanics was a theory which gave good results, for the time being, but it could not be the final theory of the submicroscopic world. "Der Herrgot würfelt nicht", "God does not play dice", as he famously said. This was the beginning of a discussion between Einstein and Niels Bohr, which lasted for many years, and did not lead to a conclusion.

In quantum mechanics, certain pairs of observables cannot be measured simultaneously with arbitrary precision. This is, for instance, the case with position and momentum. For these one has the *Heisenberg uncertainty principle* which gives an upper limit for the product of the standard deviations for position and momentum of a particle.

What quantum mechanics made rather difficult to understand was that the mathematical framework was very different from the sort of standard analysis used for classical mechanics. Instead, functional analysis, analysis in infinite dimensional spaces, developed in the beginning of the twentieth century, was needed. Using this, John von Neumann provided the mathematical foundation for quantum mechanics, with Hermann Weyl giving the group theoretical basis for the concept of symmetry.

This rigorous mathematical approach was not always appreciated by the physicists. Even most other mathematicians did not appreciate von Neumann's work. Here is an anecdote: in the thirties von Neumann gave a talk in Oxford. Afterwards G.H. Hardy, present in the audience, commented: "Yes, obviously a brilliant young man. But is this mathematics?" This is one more illustration of the sometimes irregular historical development of science.

Let us briefly describe the formalism of quantum mechanics, as used for the description of a system of N non-relativistic particles.

1. The *state* of the system is characterized by a unit vector ψ in a Hilbert space $\mathcal{H}$, in Schrödinger's viewpoint a complex-valued wave function $\psi(\mathbf{x_1}, \ldots, \mathbf{x_N})$, normalized to unity.

The term "state" comes from physics, where it means state of a physical system. Because of its importance in quantum theory, in particular in AQFT

(Algebraic Quantum Field Theory), Chapter 11, it has been taken over by mathematics, where it means a positive normalized linear functional, i.e. with $\omega(a^*a) \geq 0$ for all $a \in \mathcal{A}$ and $\omega(1_{\mathcal{A}}) = 1$, and with $\mathcal{A}$ a $*$-algebra, usually a C^*-algebra, where it has had a great career.

The collection of states on a $*$-algebra $\mathcal{A}$ is a *convex set*, i.e. for two states ω_1 and ω_2, the convex sum $\lambda_1\omega_1 + \lambda_2\omega_2$, for $\omega_1 \geq 0$, $\omega_0 \geq 0$ and $\omega_1 + \omega_2 = 1$, is again a state. If in this sum either λ_1 or λ_2 is zero, then ω is called *pure*, otherwise it is called *mixed*. A pure state is represented by a unit vector ψ in the Hilbert space $\mathcal{H}$ of a quantum mechanical system as $\omega(A) = (\psi, A\psi)$; a mixed state is described by a density operator D, according to $\omega(A) = \text{Tr}(DA)$.

A state described by a density operator is a state in quantum statistical mechanics. See the Physics intermezzo *Quantum statistical mechanics - Section 6* of this chapter.

2. *Observables* are represented by selfadjoint operators in $\mathcal{H}$. Note that most of these operators are unbounded. This is a bit of a technical problem; not very serious. All physical results can be obtained as appropriate limits from sequences of bounded operators.

3. The *interpretation* of this is as follows. Given a state described by a unit vector ψ and an observable described by the selfadjoint operator A, then the expression $(\psi, A\psi)$ is the average value of the result to be expected in a measurement. The expression $(\psi, A^2\psi)$ is the second moment of the probability distribution, and so on for all higher moments.

In this manner the pair (A, ψ) determines the probability distribution for the measurement of the observable represented by the operator A in the state described by the vector ψ.

4. The *time development* of the system is described by a 1-parameter system of unitary operators $\{U(t)\}_{t \in \mathbb{R}}$ acting on the states as $\psi(t) = U(t)\psi(0)$. $U(t)$ has the form $U(t) = e^{-\frac{i}{\hbar}Ht}$, with H the Hamiltonian operator, the energy as an observable. It is the analogue of the Hamiltonian function in the equation of motion of classical mechanics.

5. *Symmetries* are described by unitary operators $U(g)$, often representing groups $\mathcal{G}$. Important symmetries are the symmetries under spatial translations and rotations.

There are many good books on quantum mechanics. Examples are Ballentine [37], Griffiths [38], Phillips [39] and Binney et al. [40] (these four books, except the second one, complete on internet). Useful lecture notes are [41] and [42] (both on internet) and an introductory book on Hilbert space theory [43]. The items [44], [45], [46], [47] and [48] (all on internet) contain a lot of material on the Hilbert space background of the theory, much more than we can offer in this book. See however the next two mathematical intermezzos. The most important book about the mathematical foundations of quantum mechanics is von Neumann's book [49], of which the original German edition appeared in 1932. It is still of great interest. The Stanford Encyclopedia of Philosophy gives a good introduction, with attention to philosophical aspects, and a fine bibliography [50] (on internet). Articles that are basic in the history of quantum mechanics, together with commentaries of great interest are collected in a book by Barteld Leendert van der Waerden [51] (complete on internet).

- *End of physics intermezzo*

3.5   More mathematical information

- *Mathematical intermezzo*

Hilbert space. A Hilbert space $\mathcal{H}$ is a vector space with an inner product, i.e. a map $(\psi, \phi) \mapsto \mathbb{C}$, with the properties:

$(\lambda\psi, \phi) = \overline{\lambda}(\psi, \phi), \ (\psi, \lambda\phi) = \lambda(\psi, \phi)$
$(\psi_1 + \psi_2, \phi) = (\psi_1, \phi) + (\psi_2, \phi),$
$(\psi, \phi_1 + \phi_2) = (\psi, \phi_1) + (\psi, \phi_2),$
$(\psi, \psi) \geq 0$ and $(\psi, \psi) = 0$, if and only if $\psi = 0$.

All Cauchy sequences in $\mathcal{H}$, i.e. sequences with

$$\lim_{n,m\to\infty} ||\psi_m - \psi_n|| = 0$$

have limits in $\mathcal{H}$ (completeness of $\mathcal{H}$).

The inner product defines a *norm* according to $||\psi|| = \sqrt{(\psi, \psi)}$.

Problem. Show that this norm has the following properties:

$||\lambda\psi|| = |\lambda|||\psi||$, for all $\lambda \in \mathbb{C}$ and all $\psi \in \mathcal{H}$,
$||\psi_1 + \psi_2|| \leq ||\psi_1|| + ||\psi_2||$ (triangle inequality),
$||\psi|| = 0$ implies $\psi = 0$.

A inner product space $\mathcal{H}_0$, which has all above properties, except completeness is called a *pre-Hilbert space*. Adding all limits of Cauchy sequences $\{\psi_n\}_n$ gives the completion $\mathcal{H}$ of $\mathcal{H}_0$.

- *End of intermezzo*

- *Mathematical intermezzo* - **Operators in Hilbert space**. Operators are linear maps from the Hilbert space $\mathcal{H}$ into itself. The theory of operators in $\mathcal{H}$ is mathematically subtle. Operators are often only defined on a dense domain in $\mathcal{H}$. Such a domain $\mathcal{D}$ in $\mathcal{H}$ is a linear subspace such that one can obtain all of $\mathcal{H}$ as limits of convergent series in $\mathcal{D}$. Operators can be *bounded*, i.e. satisfy $||A\psi|| \leq C||\psi||$, for all elements ψ in $\mathcal{H}$, for some positive number C. The lower bound of these numbers is called the *norm* of A, denoted as $||A||$. Bounded operators defined on a dense subspace of $\mathcal{H}$ can be uniquely extended to all of $\mathcal{H}$.

Note that the notion of norm is more general than that of an inner product. This is illustrated by the case of the operator norm, which does not come from an inner product. The linear space of all bounded operators is a *normed vector space*.

A bounded operator A in $\mathcal{H}$ has a *hermitian adjoint* A^*, defined by $(A^*\psi_1, \psi_2) = (\psi_1, A\psi_2)$, for all ψ_1, ψ_2 in $\mathcal{H}$. It is *symmetric* if it satisfies $(A\psi_1, \psi_2) = (\psi_1, A\psi_2)$. For unbounded operators there is the much more subtle notion of *selfadjointness*, a key notion in the interpretation of quantum theory. A bounded operator A is called *normal* whenever $AA^* = A^*A$. For unbounded operators this remains a valid definition if one is a bit careful with the domains of definition of A and A^*.

A (bounded) operator A is *positive* if $(\psi, A\psi) \geq 0$, for all $\psi \in \mathcal{H}$. An operator U is *unitary* whenever $UU^* = U^*U = 1$. An operator P is a *projection operator* whenever $P^2 = P$ and $P^* = P$. Finite dimensional matrices have a trace, the sum of its diagonal elements, which is an invariant, with respect to change of orthonormal base.

Problem. Show that unitary and projection operators are bounded. Calculate their norms.

Unbounded operators are much harder to handle mathematically than bounded ones. Unfortunately, most of the operators in quantum theory that represent observables are unbounded. Physics books usually do not pay attention to all this. In this book this problem is at least mentioned.

Note finally that in this book all Hilbert spaces are over the complex numbers and are infinite dimensional.

The question why quantum mechanics needs Hilbert space is answered in [52] (on internet).

- *End of intermezzo*

The *spectral theorem* is central in the theory of operators in Hilbert space, and therefore in quantum theory and its physical interpretation.

- *Mathematical intermezzo.*

Spectral theorem. A hermitian (= selfadjoint) operator A in a finite dimensional Hilbert space $\mathcal{H}$ has an orthonormal basis of eigenvectors $\phi_1, \ldots \phi_n$ with real eigenvalues $\alpha_1, \ldots, \alpha_k$. Because of this it can be written as

$$A = \sum_{j=1}^{k} \alpha_j E_j,$$

with the E_j the eigenprojections associated with the eigenvectors α_j. This is the spectral theorem for the case of a finite dimensional Hilbert space, just a standard theorem in linear algebra.

For the infinite dimensional case the same is sometimes true, but very often it is not because the operator has *continuous spectrum*. In this case the spectral theorem is a highly nontrivial piece of functional analysis. In general the theorem cannot be formulated as a sum, but, in the continuous generalization of a sum, as an integral, in fact as an operator-valued Riemann-Stieltjes integral. We write it here, without explaining it, let alone proving it as

$$A = \int_{\mathbb{R}} \alpha \, dE_\alpha.$$

For a discussion of the spectral theorem, see the books of Arveson [53] and Helson [54]. Useful lecture notes are [55], [56] and [57] (all three on internet).

See for a discussion of the general Riemann-Stieltjes integral [58] and [59] (both on internet).

- *End of mathematical intermezzo*

3.6   Quantum physics 2

Quantum mechanics is the first level in quantum theory. The second is quantum statistical mechanics. To the probabilistic aspects of quantum mechanics a second layer of probability theory is added, the one that is typical of classical statistical mechanics.

- *Physics intermezzo.*

Quantum statistical mechanics. For a system of N non-relativistic particles, with N very large, there is a quantum version of classical statistical mechanics. What is 'quantized' is the algebra of observables. Note that there is no 'quantized' phase space. We have a 'non-commutative space', a heuristic algebraic notion that will be discussed in detail in Chapter 5.

We now take the observables as the primary notion, and introduce a Hilbert space $\mathcal{H}$, not as a state space as in the first case, but as an ambient background space. The *observables* are the selfadjoint elements of the (non-commutative) *-algebra $\mathcal{A}$ of all (bounded) operators in $\mathcal{H}$. Most of the observables in quantum theory are represented by unbounded operators; they appear as suitable limits of the bounded operators in $\mathcal{A}$.

1. *States* are described by *density operators* D in $\mathcal{H}$, i.e. positive selfadjoint normalized trace class operators. The operator D acts in an ambient Hilbert space $\mathcal{H}$.

In a Hilbert space the expression

$$\sum_k (e_k, D e_k)^{1/2}$$

is called the *trace* of the operator D, whenever the sum converges. In this case D is called a *trace class operator*. If the expression for the trace converges with respect to an orthonormal base, it will converge to all others.

Note that it can be proved that a bounded positive selfadjoint operator has a unique square root.

2. The observables are – again – described by selfadjoint operators in $\mathcal{H}$.

3. The *physical interpretation* is that the average value of an observable represented by the selfadjoint operator A is equal to $\mathrm{Tr}\,(AD)$, with the second moment $\mathrm{Tr}\,(A^2 D)$, etc.

4. The *time development* is described as $D(t) = e^{-\frac{i}{\hbar}Ht} D(0)\, e^{\frac{i}{\hbar}Ht}$, and the action of a *symmetry group* on the density operator as $D(g) = U(g) D\, U(g)^{-1}$.

Problem. How are symmetries represented in this scheme?

Problem. Let the density operator D be a 1-dimensional projection on the unit vector ψ. Show that the average value of an observable represented by the operator A is then equal to $(\psi, A\psi)$. This means that this case 2 is a generalization of case 1. In the first case one speaks of a *pure state*, in the second case of a *mixed state*.

Good books on quantum statistical mechanics are those of Kadanof [60] and Jancel [61], which cover both classical and quantum statistical mechanics. Useful lecture notes are [62], [63], Vilfan [64] and [65] (all four on internet). More on the density operator in [66] and [67] (both on internet). The notions of pure and mixed states are explained in [68] (on internet). A thorough treatment of the algebraic background of quantum statistical mechanics is given in the treatise (two volumes) of Brattelli et al. [69].

- End of physics intermezzo

3.7   Quantum physics 3

In this third case, *algebraic quantum theory*, again a generalization of the preceding one, there is no longer a Hilbert space at the beginning. It is primarily used in attempts to attack the mathematical problems of relativistic quantum field theory, the in a certain way very successful, but at the same time mathematically very poorly understood theory that forms the theoretical framework for the modern physics of elementary particles.

One starts with an *'abstract'* *- *algebra of observables* $\mathcal{A}$. Then there are states ω, positive normalized linear functionals on $\mathcal{A}$. In the *physical interpretation* one may construct, if one wishes, a Hilbert state space, by carrying out GNS-representation π_ω of $\mathcal{A}$, with operators $\pi_\omega(a)$, $\forall a \in \mathcal{A}$, acting in the Hilbert space $\mathcal{H}_\omega$. This GNS representation is discussed in detail in Chapter 7. The system of operators $\{\pi_\omega(a)\}_{a \in \mathcal{A}}$ is, or can be completed to the true physical algebra of observables. The 'abstract' algebra $\mathcal{A}$ should then be called the algebra of pre-observables. This will be explained later.

Time development and symmetry coincide. The theory is covariant with respect to the inhomogeneous Lorentz group. The splitting up of the

4-dimensional spacetime in separate space and time, depends on the coordinate system that is used, which means that time development has not an intrinsic meaning, So we have, strictly speaking, not an algebraic dynamical system, but what was called in Chapter 1, an *algebraic covariance system*. The algebra $\mathcal{A}$ is composed of algebras $\mathcal{A}(O)$, for each open set O in spacetime. All this will be discussed in detail in Chapter 7.

3.8 Concluding remarks

In this chapter we have shown that one may distinguish three levels in quantum theory, each a generalization of its predecessor.

Level 1. *Quantum mechanics*. A Hilbert space $\mathcal{H}$ of pure state vectors ψ, observables selfadjoint operators A in $\mathcal{H}$.

Level 2. *Quantum statistical mechanics*. A concrete algebra of observables $\mathcal{A}$, operators in an ambient Hilbert space $\mathcal{H}$. States as density operators D acting on $\mathcal{A}$.

Level 3. *Algebraic quantum field theory*. An abstract algebra $\mathcal{A}$ of observables, with as states general positive normed linear functionals on $\mathcal{A}$. Important role of the GNS representation construction.

Included in this last case are non-relativistic examples such as infinite quantum spin systems, in the thermodynamic limit. Translation and rotation symmetry may then be important.

References

1. Goldstein, H., Poole, Ch., Safko, J.: Classical Mechanics. Third edition. Addison Wesley 2001. The complete book is available at
   http://www.cmi.ac.in/~souvik/books/mech/Goldstein.pdf
   or
   http://poincare.matf.bg.ac.rs/~zarkom/
   Book_Mechanics_Goldstein_
   Classical_Mechanics_optimized.pdf
2. Kibble, T., Berkshire, F.H.: Classical Mechanics. Fifth edition. Imperial College press 2004.
3. Taylor, J.R.: Classical Mechanics. University Science Books 2005. The complete book is available at
   http://dl1.ponato.com/eb1/760__0a09e60.pdf
4. Sparks, J.: Classical Mechanics. Lectures university of Oxford 2015. Available at
   http://users.ox.ac.uk/~math0391/CMlectures.pdf
5. Orfanidis, S.J.: Maxwell's Equations. Rutgers University lectures. No year. Available at

http://www.ece.rutgers.edu/~orfanidi/ewa/ch01.pdf

6. Gross, P.W., Kotiuga, P.R.: Electromagnetic Theory and Computation. A topological Approach. Cambridge University Press 2004. The complete book is available at

http://library.msri.org/books/Book48/
files/gross-kotiuga.pdf

7. Dyson, F.J.: Why is Maxwell's Theory so hard to understand? Available at

http://www.damtp.cam.ac.uk/user/tong/em/dyson.pdf

8. Gibbs, J.W.: Elementary Principles in statistical mechanics. Original edition Scribner 1902, Reprinted by Dover 2014. The complete 1902 edition can be obtained at

http://myrtille.ujf-grenoble.fr/pagesperso/
bahram/Phys_Stat/Biblio/gibbs_1902.pdf

9. Khinchin, A.Y.: Mathematical Foundations of Statistical Mechanics. Dover 1949.

10. Huang, K.: Statistical Mechanics. Second Edition. Wiley 1987. The complete book is available at

http://www.fulviofrisone.com/attachments/article/
485/Huang%20-%20Introduction%20to%20Statistical%20
Physics,%20Taylor%20and%20Francis,%202001_305.pdf

11. Van Vliet, C.M.: Equilibrium and Non-equilibrium Statistical Mechanics. Revised Edition. World Scientific 2008.

12. Gallavotti, G.: Statistical Mechanics. Lecture notes University of Rome, La Sapienza 1998. Available at

http://www.pd.infn.it/~saggion/libro.pdf

13. Fitzpatrick, R.: Thermodynamics & Statistical Mechanics. Lecture notes University of Texas at Austin. No year. Available at

http://farside.ph.utexas.edu/teaching/sm1/statmech.pdf

14. Frigg, R.: What is Statistical Mechanics? London School of Economics Lectures. No year. Accessible at

http://www.romanfrigg.org/writings/What_is_SM.pdf

15. Huan, A.: Statistical Mechanics. Lectures Nanyang Technological University. No date. Available at

http://www.spms.ntu.edu.sg/PAP/courseware/statmech.pdf

16. Dobrushin, R.L.: A Mathematical Approach to Foundations of Statistical Mechanics. Preprint Vienna Schrödinger Institute for Mathematical Physics 1994. Available at

http://www.mat.univie.ac.at/~esiprpr/esi179.pdf

17. Powers, J.P.: Lectures on Thermodynamics. University of Notre Dame lectures notes 2016. Available at

https://www3.nd.edu/~powers/ame.20231/notes.pdf

18. NN. Radon Measures. University of North Carolina Lectures. No date. Available at

http://www.unc.edu/math/Faculty/met/measch13.pdf

19. Heil, C.E.: Borel and Radon Measures on the Real Line. Georgia

Institute of Technology Lecture Notes. No year. Available at
http://people.math.gatech.edu/~heil/
handouts/appendixd.pdf

20. Arveson, W.: Notes on Measure and Integration in Locally Compact Spaces. Lecture notes University of California at Berkeley 1996. Available at
https://math.berkeley.edu/~arveson/Dvi/rieszMarkov.pdf

21. Halmos, P.R.: Measure Theory. Corrected printing. Springer 1978.

22. Gupta, M.R.: A Measure Theory Tutorial. Course at the Department of Electrical Engineering, University of Washington at Seattle 2006-2008. Available at
https://www.ee.washington.edu/techsite/
papers/documents/UWEETR-2006-0008.pdf

23. Tao, T.: An Introduction to Measure Theory. American Mathematical Society 2011. A preliminary version is available at
https://terrytao.files.wordpress.com/2012/12/
gsm-126-tao5-measure-book.pdf

24. Hunter, J.K. : The Riemann Integral. University of California at Davis lectures. No year. Available at
https://www.math.ucdavis.edu/~hunter/m125b/ch1.pdf

25. Zitkovic, G.: The Lebesque integral. University of Texas lecture notes. 2013. Available at
https://www.ma.utexas.edu/users/gordanz/
notes/lebesgue_integration.pdf

26. Melrose, R.: The Lebesgue Integral. Lecture Notes. MIT Department of Mathematics. No year. Available at
https://math.mit.edu/~rbm/18-102-S14/Chapter2.pdf

27. Anewski, D.: Riemann-Stieltjes integrals. Lund University lecture notes. 2012. Available at
http://www.maths.lth.se/matstat/kurser/mas213/
riemannstieltjes.pdf

28. Geller, D.N.: Lebesgue-Stieltjes integrals. SUNY Stony Brook lectures. No year. Available at
https://www.math.stonybrook.edu/~daryl/ls.pdf

29. Kolmogorov, A.N.: Foundations of the Theory of Probability. Second edition. Chelsea 1960. Available at
http://www.york.ac.uk/depts/maths/histstat/
kolmogorov_foundations.pdf

30. Chung, K.L.: A Course in Probability Theory. Third Edition. Academic Press 2000.

31. Klemke. A.: Probability Theory : A Comprehensive Course. Second Edition. Springer 2014.

32. Dudley, R.M.: Real Analysis and Probability Theory. Cambridge University Press 2002.

33. Morey, E.R.: The basics of probability theory. Lecture notes University of Colorado at Boulder 2010. Available at

`http://www.colorado.edu/economics/`
`morey/7818/probtheory/basicsofprobabilitytheory.pdf`
34. Durrett, R.: Probability : Theory and Examples. Cambridge University Press 2010. Available at
`https://services.math.duke.edu/~rtd/PTE/PTE4_1.pdf`
35. Knills, O.: Probability Theory and Stochastic Processes with Applications. Overseas Press 2009. Available at
`http://www.math.harvard.edu/~knill/books/`
`KnillProbability.pdf`
36. Löwe, M.: Probability Theory. Lecture notes Radboud University Nijmegen 2001-2002. Available at
`http://www.math.ru.nl/~hendriks/ProbabilityC.pdf`
37. Ballentine, L.E.: Quantum Mechanics: A Modern Development. World Scientific 1998. The complete 673 page book is available at
`http://www-dft.ts.infn.it/~resta/fismat/ballentine.pdf`
38. Griffith, D.: Introduction to Quantum Mechanics. Second edition, Prentice Hall 1995. 39. Phillips, A.C.: Introduction to Quantum Mechanics. Wiley 2003. The complete book is available at
`http://www.fisica.net/quantica/`
`Phillips%20-%20Introduction`
`%20to%20Quantum%20Mechanics.pdf`
40. Binney, J., Skinner, D.: The Physics of Quantum Mechanics. Oxford University Press 2013. The complete 310 page book available at
`http://www-thphys.physics.ox.ac.uk/people/`
`JamesBinney/qb.pdf`
41. Morin, D.J.: Introduction to Quantum Mechanics. Lecture notes Harvard University. No year. Available at
`http://www.people.fas.harvard.edu/~djmorin/`
`waves/quantum.pdf`
42. Timoney, R.: Hilbert Spaces. Lecture notes Trinity College, Dublin 2008-2009. Available at
`http://www.maths.tcd.ie/~richardt/321/321-ch4.pdf`
43. Alabiso, C., Weiss, I.: A Primer on Hilbert Space Theory. Springer 2015.
44. Avramidi, I.: Notes on Hilbert Spaces. Lectures New Mexico Tech 2000. Available at
`http://infohost.nmt.edu/~iavramid/notes/hilbert.pdf`
45. Hunter, K.J., Nachtergaele, B.: Hilbert Spaces. Chapter 6 of Applied Analysis. World Scientific 2000. Chapter 6 is available at
`https://www.math.ucdavis.edu/~hunter/book/ch6.pdf`
The complete book at
`http://www.math.ucdavis.edu/~hunter/book/`
46. Landsman, N.P.: Hilbert Spaces and Quantum Mechanics. Lectures Radboud University Nijmegen University 2006. Available at
`http://www.math.kun.nl/~landsman/HSQM2006.pdf`
47. Elster, Ch.: Quantum Mechanics in Hilbert Spaces. Lecture notes Ohio

University. No year. Available at
http://www.phy.ohiou.edu/~elster/lectures/qm1_1p2.pdf
48. Teschl, G.: Mathematical Methods in Quantum Mechanics. Am. Math. Soc. 2009. The complete book can be found at
http://www.mat.univie.ac.at/~gerald/
ftp/book-schroe/schroe.pdf
49. von Neumann, J.: Mathematical Foundations of Quantum Mechanics. Translated from the German. Princeton 1955,
50. Jennan, I.: Quantum Mechanics. Stanford Encyclopedia of Philosophy 2015. Accessible at
http://plato.stanford.edu/entries/qm/
51. Van der Waerden, B.L.: Sources of Quantum Mechanics. Dover 2007. Available at
https://archive.org/details/SourcesOfQuantumMechanics
52. Scrinzi, A.: Why we do quantum mechanics in Hilbert space. Lecture notes University of München. 2012. Available at
http://www.mathematik.uni-muenchen.de/~lerdos/
WS12/MQM/mathQM_hilbert.pdf
53. Arveson, W.: A Short Course on Spectral Theory. Springer 2002.
54. Kowalski, E.: Spectral theory in Hilbert spaces. Lecture notes ETH Zürich. No year. Available at
https://people.math.ethz.ch/~kowalski/
spectral-theory.pdf
55. Taylor, M.: The Spectral Theorem for Self-Adjoint and Unitary Operators. Lecture notes University of North Carolina at Chapel Hill. No year. Available at
http://www.unc.edu/math/Faculty/met/specthm.pdf
56. Garrett, P.: Eigenvectors, spectral theorems. Lecture notes University of Minnesota. No year. Available at
http://www.math.umn.edu/~garrett/m/
algebra/notes/24.pdf
57. Helson, H.: The Spectral Theorem. Springer 1986.
58. Anevski, D.: Riemann-Stieltjes integrals. Lund University lecture notes 2012. Available at
http://www.maths.lth.se/matstat/kurser/
mas213/riemannstieltjes.pdf
59. Leffler, K.: The Riemann-Stieltjes integral. Umea University lecture notes 2014. Available at
http://umu.diva-portal.org/smash/get
diva2:719488/FULLTEXT01.pdf
60. Kadanov, L.P., Baym, G., Pines, D.: Quantum Statistical Mechanics. Westview Press 1994.
61. Jancel, R.: Foundations of Classical and Quantum Statistical Mechanics. Pergamon Press 1969.
62. Fendley, P.: Quantum statistical mechanics from classical mechanics. Lecture notes Oxford university 2015. Available at

`http://users.ox.ac.uk/~phys1116/msm2.pdf`
These notes are part of a full course on Modern Statistical Mechanics. Available at
`http://users.ox.ac.uk/~phys1116/`

63. Garanin, D.: Statistical Thermodynamics. Lecture notes City University of New York 2012. Available at
`http://www.lehman.edu/faculty/dgaranin/`
`Statistical_Thermodynamics/Statistical_physics.pdf`

64. Vilfan, J.: Lecture Notes in Statistical Mechanics. Lectures at the Abdus Salam Institute, Trieste. No year. Available at
`http://www-f1.ijs.si/~vilfan/SM/`

65. Modo, M.: Quantum Statistical Mechanics. 2012. Available at
`http://physics.unifr.ch/admin/dbproxy.php?`
`table=fuman_filepool&column=content&id=1209`

66. Soper, D.E.: The density operator in quantum mechanics Lecture notes university of Oregon 2012. Available at
`http://pages.uoregon.edu/soper/`
`QuantumMechanics/density.pdf`

67. Bertlmann, R.A.: Density Matrices. Notes University of Vienna. No year. Available at
`http://homepage.univie.ac.at/reinhold.bertlmann/`
`pdfs/T2_Skript_Ch_9corr.pdf`

68. Van Enk, S.J.: Mixed states and pure states. Lecture notes university of Oregon. 2009. Available at
`http://pages.uoregon.edu/svanenk/`
`solutions/Mixed_states.pdf`

69. Brattelli, O., Robinson, D.W.: Operator Algebras and Quantum Statistical Mechanics. 1. C^*- and W^*-Algebras. Symmetry Groups. Decomposition of States. 2. Models in Quantum Statistical Mechanics. Springer 2002, 2003.

4 NON-COMMUTATIVE TOPOLOGY AND GEOMETRY

4.1 Historical introduction

The idea of describing geometric concepts in algebraic terms is an old one. In its mature form it was independently invented in the seventeenth century by René Descartes and Pierre Fermat.

A quote as illustration : "Et je craindrai pas d'introduire ces termes d'arithmétique en la géométrie, afin de me rendre plus intelligible." (René Descartes, La Géométrie, 1638).

English translation: "I would not be afraid to introduce these termes of arithmetic in geometry in order to make myself better understood."

Nowadays it is a subject that is taught as *analytical geometry* in secondary schools. Its generalized form, formulated in arbitrary finite dimensions, is *linear algebra*, a topic in the first year of the mathematics curriculum at the university, with *functional analysis*, its infinite dimensional generalization, as a topic for graduate courses.

4.2 The modern situation

More recent is 'algebraization of topology'. This started with the 1943 paper by Israel Gelfand and Mark Aronovich Naimark [1]. It inspired a whole new and important field of mathematical research: non-commutative topology and geometry, the subject of this chapter. For the general importance of this paper see the review by Alain Connes [2]. There is also *algebraic geometry*, an important branch of mathematics, in which one studies *algebraic varieties*, the null sets of systems of complex polynomials. We shall not discuss this.

- Mathematical intermezzo - extra

Topology, topological spaces. A basic subject in mathematics; as such it deserves an intermezzo.

Topology describes surfaces in a manner in which only overall form counts. In $\mathbb{R}^3$ a sphere is topologically the same as a cube, but different from a torus. To distinguish between spheres and cubes one needs *differential geometry, differentiable manifolds.* See the mathematical intermezzo on this topic in Section 5 of this chapter.

The definition of a topological space is based on the notions of open and closed set. Before discussing topological spaces it is therefore useful to give first a bit of information on general set theory.
Properties of and relations between sets.

1. Inclusion: $A \subset B$, meaning $x \in A \rightarrow x \in B$.
The converse is $B \supset A$.
2. Complement: A^c, $x \in A^c \rightarrow x \notin A$.
3. Union: $A \cup B$, meaning $x \in A \cup B$ i.e. $x \in A$ *or* $x \in B$.
4. Intersection: $A \cap B$, meaning $x \in A$ *and* $x \in B$.
Note that union and intersection are meaningful for systems of sets, possibly infinite.

Problem. Prove the following identities:

$$(A \cup B)^c = A^c \cap B^c, \quad (A \cap B)^c = A^c \cup B^c.$$

The book by Williams [3] (complete on internet) gives a good introduction to set theory.

The formal definition of a topological space is as follows:

A *topological space* is a pair $(X, \mathcal{T})$, consisting of a nonempty set X and a *topology*, a system $\mathcal{T}$ of subsets of X, the *open sets*, which satisfies the following convenient system of axioms:

1. X and the empty set $\emptyset$ belong to $\mathcal{T}$,
2. The union $\cup_\alpha A_\alpha$ of an *arbitrary* system $\{A_\alpha\}_\alpha$ of sets in $\mathcal{T}$ is again in $\mathcal{T}$.
3. The intersection $\cap_\alpha A_\alpha$ of a *finite* system $\{A_\alpha\}_\alpha$ of sets in $\mathcal{T}$ is again in $\mathcal{T}$.

A set is a *closed set* if and only if it is the complement of an open set.

Problem. A topological space can be equivalently defined in terms of closed sets. Give the system of axioms for this definition.

It is clear that X and the empty set $\emptyset$ are open as well as closed. If more pairs of sets in $\mathcal{T}$ have this property, then A and A^c are obviously disjoint. If there are one or more of such pairs of sets, the topological space consists of a number of disjoint topological spaces; it is said to be *disconnected*. A connected topological space is called *simply connected* if and only if every closed curve can be continuously contracted to a point. A sphere is simply connected; a torus is not.

The very important notions of continuity and limit are based on the presence of a topology.

Neighbourhood. A neighbourhood of a point x in a topological space $(X, \mathcal{T})$ is a subset V of X for which there exists an open subset U of X, such that $x \in U \subset V$.

Continuity. A map f from a topological space $(X_1, \mathcal{T}_1)$ to a topological space $(X_2, \mathcal{T}_2)$ is continuous in a point x_1 in X_1 if and only if every inverse image of a neighbourhood of $f(x_1)$ is a neighbourhood of x_1 in X_1. One similarly defines continuity on an open set in X_1 or on all of X_1.

The real line $\mathbb{R}^1$ has a standard topology in which all open sets are the arbitrary unions of open intervals (a, b).

Problem. Show that in this case a real-valued function f is continuous in a point x_1 whenever for every $\epsilon > 0$ there is a $\delta > 0$ such that for $x \in (x_1 - \delta, x_1 + \delta)$ one has $f(x) \in (f(x_1) - \epsilon, f(x_1) + \epsilon)$.

Convergence. An infinite sequence of points $x_1, x_2, \ldots$ is said to have as limit the point x if and only if there is for every neighbourhood U_x an N such that for every $n \geq N$ one has $x_n \in U_x$.

In a *Hausdorff topological space* each pair of distinct points have disjoint neighbourhoods. This ensures that a convergent sequence has a unique limit. All the topological spaces in this book are supposed to be Hausdorff.

Compactness. A topological space X, or a subset of X, is called compact if and only if every covering by open sets has a finite subcover. Compactness is an important 'tightness' property of topological spaces. All sorts of properties are usually simpler to prove in compact spaces. A subset of $\mathbb{R}^n$ is compact if and only if it is closed and bounded. $\mathbb{R}^n$ as a whole is locally compact, a slight generalization of compactness, which we shall not discuss here.

Topology can be divided in two subtopics: 1. Point set topology, describes local notions, with points and neighbourhoods, to be used for defining limits and continuity. 2. Algebraic topology, describes global properties, for instance the difference between a sphere and a torus. In this book the main role is played by notion 1.

A surface in general topology is an 'elastic' object. Under deformation it remains the same as long as it is not torn. It also has no smoothness properties. An n-dimensional sphere and an n-dimensional cube are the same objects seen as topological spaces.

A general topological space has no metric properties. There is however a special class of topological spaces which do have a metric.

A *metric space* is a set X with a *distance function* $d(\cdot, \cdot)$, with the properties:

1. $d(x, y) \geq 0$,
2. $d(x, y) = 0 \iff x = y$,
3. $d(x, y) = d(y, x$ (symmetry),
4. $d(x, z) \leq d(x, y) + d(y, z)$ (triangle inequality),

for all x, y, z in X.

One defines a system of open sets in a metric space: a subset U of a metric space X is open whenever it can be written as a union of open balls $\{x \in X \mid d(x, x_n) < \alpha$, with $\alpha > 0$ and all points $x_n \in X$.

A metric space is in this manner indeed a topological space, with a metric which defines its topology, such that, for instance, convergence of sequences $\{x_n\}_n$ to a point x_0 means $\lim_{n \to \infty} d(x_n, x_0) = 0$. A topological space is *metrizable* whenever its topology can be defined by a metric. This metric need not be unique or explicitly given. A metric or metrizable space is *complete* if it contains the limit points of all convergent sequences.

For good introductions in topology, see the books of Mark Armstrong [4] and Karl Seifert with William Threlfall [5] (complete on internet). For a discussion on compactness and a definition of the Hausdorff property, see [6].

- *End of mathematical intermezzo*

Gelfand and Naimark formulated in their 1943 paper a theorem, the first part of which says that the continuous functions on a compact topological space form a commutative C^*-algebra.

See - Mathematical intermezzos *Algebras* - to be found in Chapter 2, Section 2 and C^*-*algebras* - to be found in Chapter 6, Section 2.

The second more surprising part states that any commutative C^*-algebra with a unit element is algebraically isomorphic to the algebra of continuous functions $C(X)$ on a compact topological space X, which means that there is a one-to-one correspondence – up to obvious equivalence – between the above 'explicit' function algebras and 'abstract' commutative C^*-algebras. Moreover, starting from a given commutative C^*-algebra, the corresponding function algebra can be reconstructed. (We leave apart a slight generalization, in which X may be only locally compact and in which a unit element in $\mathcal{A}$ is not needed.)

The above may be illustrated by the following picture.

compact topological space X

$\updownarrow$

algebra of continuous functions $C(X)$

$$\updownarrow$$

commutative C^*-algebra

4.3 Algebraization of topology

All this suggests that a topological space can be studied algebraically. For algebraic topology in the above sense there is a well-established field, called – not surprisingly – algebraic topology, which indeed uses tools from abstract algebra to study topological spaces. See lecture notes May [7] (on internet) and the book by Allen Hatcher [8] (complete on internet).

4.4 A Gelfand-Naimark theorem for manifolds

Is there a Gelfand-Naimark theorem for differential manifolds? Can a manifold $\mathcal{M}$ be characterized by the commutative algebra $C^\infty(\mathcal{M})$ of its smooth functions?

The answer is affirmative. A remarkable theorem was proved by Erik Thomas [9], stating that if – under a very weak restriction, two manifolds $\mathcal{M}_1$ and $\mathcal{M}_2$ are diffeomorphic,Â i.e. the same as manifolds, then the algebras $C^\infty(\mathcal{M}_1)$ and $C^\infty(\mathcal{M}_2)$ are algebraically isomorphic. Furthermore, if $C^\infty(\mathcal{M})$ is given as an 'abstract' algebra, $\mathcal{M}$ can be reconstructed from it.

The restriction on the theorem is that the manifolds should be σ-compact, a notion that we will not explain here. Almost all manifolds in physics are σ-compact.

In the comparison with the Gelfand-Naimark theorem for topological spaces one thing is missing: there is no clear characterization of the type of 'abstract' algebra that is involved. It may be some kind of Fréchet algebra. See for this notion the Mathematical intermezzo - Fréchet algebras, found in Chapter 6, Section 5. An attempt in this direction was made by Fatima Azmi [10].

We may conclude that we have a picture that is very similar to a part of the scheme for topological spaces:

manifold $\mathcal{M}$

$\updownarrow$

algebra of functions $C^\infty(\mathcal{M})$

$\downarrow$

all geometric objects: vector fields, differential forms, tensor fields, etc.

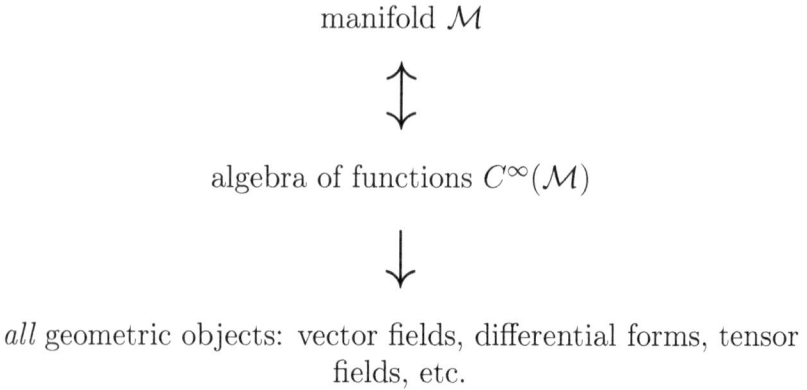

This diagram expresses the fact that the geometry of a manifold $\mathcal{M}$ can be based on a commutative algebra, the algebra $C^\infty(\mathcal{M})$ of its smooth, i.e. infinitely differentiable functions. One might therefore use the terms "commutative differential geometry" and "commutative manifold". It gives a suggestion for a generalization to "non-commutative differential geometry" and "non-commutative manifold", as we shall see further on.

- Mathematical intermezzo

Differentiable manifolds, differential geometry. A differentiable or smooth manifold, or just a manifold, is a topological space with additional structure that allows differentiation of functions on it. Originally a manifold of dimension n was defined as embedded in some $\mathbb{R}^{n+1}$. Think of the surface of a 2-sphere, as embedded in $\mathbb{R}^3$. The modern approach sees an n-dimensional manifold as an object in its own right, made up from overlapping images of open pieces of $\mathbb{R}^n$, *charts*, connected by differentiable transition functions. Analysis on differentiable manifolds is called *differential geometry*.

On a manifold $\mathcal{M}$ one has *vector fields*, smooth assignments of tangent vectors X_p at each point p of $\mathcal{M}$. A dual notion is that of *(differential) 1-forms*, smooth assignments of vectors X_p^* from the dual of the tangent spaces at each p. Differential forms are very important tools in differential geometry. In addition to 1-forms, there are 0-forms, just smooth functions on $\mathcal{M}$, and then, less trivially, a sequence of k-forms for k from 2 to n. They are connected by the *exterior derivative* d, a linear map from k-forms into $(k+1)$-forms, with the property $d^2 = 0$.

Differential manifolds may have a metric structure given by a *metric tensor*. It is then a *Riemannian manifold*. Such manifolds provide the mathematical description of spacetime in general relativity. There are many good books on differentiable manifolds and differential geometry. Examples are the books by Warner [11] and Lang [12] and also Lee [13] and Kosinski [14] (both on internet - complete). Useful lecture notes are [15] and [16] (both on internet).

- End of mathematical intermezzo

A consequence of the fact that a manifold $\mathcal{M}$ is completely characterized by its algebra of smooth functions $C^\infty(\mathcal{M})$ is that all sorts of properties of $\mathcal{M}$ can be described in terms of algebraic properties of $C^\infty(\mathcal{M})$. The above intermezzo gave a geometrical definition of a vector field X on $\mathcal{M}$. Algebras $\mathcal{A}$ have *derivations*, i.e. linear maps $D : \mathcal{A} \mapsto \mathcal{A}$, characterized by the relation $D(ab) = (Da)b + a(Db)$, for all a and b in $\mathcal{A}$. A well-known theorem states that there is a one-to-one correspondence between vector fields on $\mathcal{M}$ and derivations on $C^\infty(\mathcal{M})$. In this manner a vector field becomes a purely algebraic object. The same is true for differential forms, general tensor fields, etc., as is indicated in the picture. Note that in this description the points of $\mathcal{M}$ have disappeared from sight. It is good to remember this in the following sections where we shall deal with *virtual 'spaces'*.

4.5   A non-commutative generalization

4.5.1   The case of topological spaces

We have seen that important notions for topological spaces have translations to notions for the commutative algebra that characterize these spaces. It is not surprising that at a certain moment the question arises what meaning these notions have in related non-commutative algebras, for instance non-commutative instead of commutative C^*-algebras. An intuitive and heuristic answer is that they describe the properties of 'non-commutative topological spaces' with the non-commutative algebras, again heuristically speaking, being 'non-commutative algebras of functions'.

One may think of a scene in Alice in Wonderland. Alice watches a cat on the branch of a tree. The cat smiles at her. At a certain moment the cat starts slowly to become vaguer and then disappears completely. But its grin remains. The smile of a virtual cat.

Of course, in a mathematically rigorous manner this does not make sense, but it is a fruitful intuitive idea which suggests interesting properties and operations in the non-commutative algebra, mimicking the properties in the original commutative algebra.

As a result one obtains a rigorous notion of *noncommutative algebraic topology*, something which has been fruitfully studied and developed by many authors.

4.5.2 The case of manifolds

The next step is to develop 'non-commutative differential geometry'.

The diagram expresses the fact that the geometry of a manifold $\mathcal{M}$ can be based on a commutative algebra, the algebra $C^\infty(\mathcal{M})$ of its smooth, i.e. infinitely differentiable functions. One might therefore use the terms commutative differential geometry and commutative manifold. It gives a suggestion for a generalization to non-commutative differential geometry and non-commutative manifold.

For this consider, besides of the commutative algebra $\mathcal{A} = C^\infty(\mathcal{M})$, an algebra $\widehat{\mathcal{A}}$, non-commutative, but in some way or another similar to $\mathcal{A}$, for instance obtained from $C^\infty(\mathcal{M})$ by a deformation procedure.

An extended diagram illustrates the basic idea of non-commutative geometry:

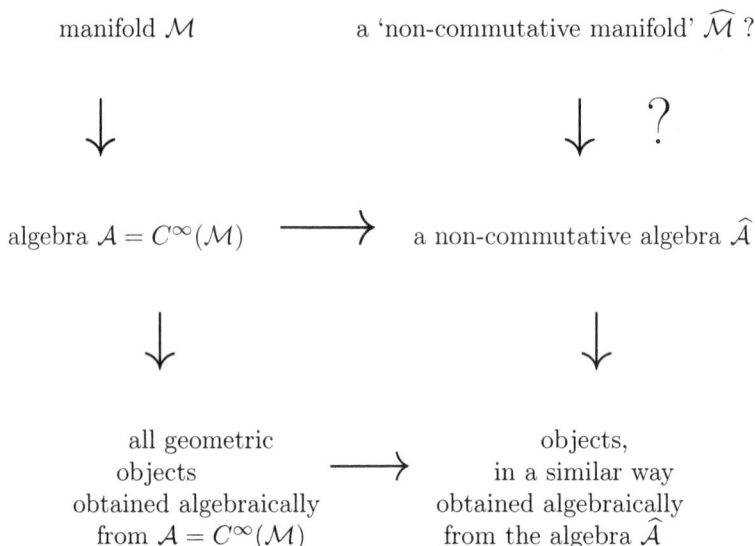

$$
\begin{array}{ccc}
\text{manifold } \mathcal{M} & & \text{a 'non-commutative manifold' } \widehat{\mathcal{M}} \text{ ?} \\[1em]
\downarrow & & \downarrow \ ? \\[1em]
\text{algebra } \mathcal{A} = C^\infty(\mathcal{M}) & \longrightarrow & \text{a non-commutative algebra } \widehat{\mathcal{A}} \\[1em]
\downarrow & & \downarrow \\[1em]
\begin{array}{c}\text{all geometric}\\ \text{objects}\\ \text{obtained algebraically}\\ \text{from } \mathcal{A} = C^\infty(\mathcal{M})\end{array} & \longrightarrow & \begin{array}{c}\text{objects,}\\ \text{in a similar way}\\ \text{obtained algebraically}\\ \text{from the algebra } \widehat{\mathcal{A}}\end{array}
\end{array}
$$

In the right-hand column one mimicks the various definitions, based on $\mathcal{A} = C^\infty(\mathcal{M})$ in the left-hand side. This turns out to be a fruitful idea; many of the ordinary geometric definitions still make sense in the non-commutative algebraic context.

The typical example of this is the notion of vector field. A *vector field* on a manifold $\mathcal{M}$ is a smooth assignment of a vector space T_p, the tangent space at p, to each point p of $\mathcal{M}$. A *derivation* of an algebra $\mathcal{A}$ (commutative or non-commutative) is a linear map $D : \mathcal{A} \longrightarrow \mathcal{A}$, with the property

$$D(ab) = (Da)b + a(Db), \quad \forall\, a, b \in \mathcal{A}.$$

For the algebra $C^\infty(\mathcal{M})$ of smooth vector fields on $\mathcal{M}$ the vector fields on $\mathcal{M}$ are precisely the derivations on $C^\infty(\mathcal{M})$.

For a non-commutative algebra the notion of derivation is well-defined. A derivation on $\widehat{\mathcal{A}}$ can then be seen as a 'non-commutative vector field' on the (non-existing) 'non-commutive manifold $\widehat{\mathcal{M}}$'.

In this manner almost all geometric notions can be defined in the context of non-commutative algebras.

Conclusion. *Strictly speaking, 'non-commutative manifolds' do not exist; it is an imaginary but nevertheless very fruitful idea which gives intuitive guidance for interesting work on certain algebra structures which generalizes those from ordinary differential geometry.*

The person behind this complex of ideas is the French mathematician Alain Connes, Fields Medalist in 1982 for his work on operator algebras. He is the creator of what is now generally known as *non-commutative geometry*. See his book [17] (complete on internet). It is not exactly easy reading, but very stimulating. Irving Segal wrote a long, interesting and mostly positive review [18] (on internet), in which he called it a "a long discourse or letter to friends". Connes was not amused. A standard textbook on the subject has been written by Jose Gracia-Bondía, Joseph Várilly and José Figueroa [19]. There are introductory lectures by Fabien Besnard [20], Peter Bongaarts [21] and Andrzej Sitarz [22] (all three available on internet).

Connes applied the idea of a 'non-commutative space' to Riemannian manifolds, a particular kind of differentiable manifold.

Connes is one of the few mathematicians who is interested in modern physics, and moreover has a broad knowledge of it. One of his ideas has been to apply non-commutative geometry to elementary particle physics, in particular to the problems of relativistic quantum field theory, the main theoretical basis of particle theory. See Chapter 8, Section 2. This theory is plagued by divergence problems. They appear in particular at very short distances in spacetime, a 4-dimensional pseudo-Riemannian manifold. (Pseudo in this context means that the metric is indefinite, not positive definite, as in a standard Riemannian manifold). The idea of Connes was to use a 'non-commutative Riemannian manifold' for spacetime, to remove these divergences. See for a paper of Connes explaining his aims in applying non-commutative geometry to physics [23] (on internet). All this led him to a beautiful theory of general non-commutative Riemannian manifolds, even though the consequences for the divergence problems of quantum field theory remain disappointing.

References

1. Gelfand I.M., Naimark, M.A.: *On the imbedding of normed rings into the ring of operators on a Hilbert space.* Math. Sbornik 12, 197-217 (1943).
2. Connes, A. : On the Foundation of Noncommutative geometry. No year. Available at
http://www.alainconnes.org/docs/gelfand.pdf
3. Williams, A.R.: An Introduction to Set Theory. Create Space Independent Publishing Platform 2014. The book is available at
http://www.math.toronto.edu/weiss/set_theory.pdf
4. Armstrong, M.A.: *Basic Topology.* Springer 1983 1997.
5. Seifert, K., Threlfall, H.W.: *A Textbook of Topology.* Academic Press 1980. The complete book can be found at
http://www.maths.ed.ac.uk/~aar/papers/seifthreng.pdf
6. Jackson, B.: Hausdorff Spaces and Compact Spaces. Queen Mary University of London Lecture Notes. No year. Available at
http://www.maths.qmul.ac.uk/~bill/topchapter3.pdf
7. May, J.P.: *A Concise Course in Algebraic Topology.* Mathematics Department University of Chicago Lecture Notes. 2014-2015. Accessible at
http://www.math.uchicago.edu/
~may/CONCISE/ConciseRevised.pdf
8. Hatcher, A. *Algebraic Topology.* 2001. The complete book is accessible at
https://www.math.cornell.edu/~hatcher/AT/AT.pdf
9. Thomas, E.: *Characterization of a Manifold by the $*$-algebra of its C^∞-functions.* University of Groningen preprint, unpublished. Erik Thomas passed away in 2011. Scans of this preprint can be obtained from the author of this book. (p.j.m.bongaarts[at]xs4all.nl)
10. Azmi, F.: *Differential Fréchet $*$-algebras and characterization of smooth functions on* $\mathbb{R}$. Rocky Mountain J. of Math. 42, 1777-1786 (2012)
11. Warner, F.: *Foundations of Differentiable Manifolds and Lie Groups.* Springer 1983
12. Lang, S.: *Introduction to Differential Manifolds, Second Edition.* Interscience 1962. The complete book can be accessed at
http://im0.p.lodz.pl/~kubarski/AnalizaIV/Wyklady/
Lang%20S.%20Introduction%20to%20Differentiable%
20Manifolds%20(ISBN%200387954775)
(Springer,%202002)(263s)_MDdg_.pdf
13. Lee, J.M.: *Introduction to Smooth Manifolds.* Springer 2002. The complete book is accessible at
http://www.math.ku.dk/~moller/e01/speciel/smooth.pdf
14. Kosinski, A.A.: Differential Manifolds. Dover 2007. Available at
http://www.maths.ed.ac.uk/~aar/papers/kosinski.pdf
15. Fernandes, R.L.: *Differential Geometry.* Lectures at the Mathematics Department of the University of Illinois at Urbana-Champaign 2013. No year. Accessible at

```
http://www.math.illinois.edu/~ruiloja/
Math518/notes.pdf
```
16. Hitchin, N.: *Differential Manifolds*. Lectures at the Mathematical Institute, Oxford University. 2012. Accessible at
```
https://www.ime.usp.br/~gorodski/teaching/
mat5799-2015/hitchin-manifolds2012.pdf
```
17. Connes, A.: *Noncommutative Geometry*. Academic Press 1994. The book is out of print, but can be downloaded from Connes' website
```
http://www.alainconnes.org/docs/book94bigpdf.pdf
```
18. Segal, I.E.: *Noncommutative geometry, by Alain Connes*. Bull. Amer. Math. Soc. 33, 459-465 (1995). Can be found at
```
http://www.ams.org/bull/1996-33-04/
S0273-0979-96-00687-8/S0273-0979-96-00687-8.pdf
```
19. Gracía-Bondia, J.M., Várilly, J.C., Figueroa, H. : *Elements of Noncommutative Geometry*. Birkhäuser 2001.
20. Besnard, F. *Friendly introduction to the concepts of non-commutative geometry*. Notes for the seminar "Philosophie et Physique". Paris 2013. Available at
```
http://fabien.besnard.pagesperso-orange.fr/
cours/notesfriendly.pdf
```
21. Bongaarts, P.: *A short introduction to noncommutative geo-metry*. Talk given at the Institute Lorentz for Theoretical Physics, Leiden 2004. Available at
```
https://www.lorentz.leidenuniv.nl/
modphys/ncg-lecture.pdf
```
22. Sitarz, A. : *A friendly introduction to Noncommutative Geometry*. Lectures given at the Zakopane Summer school, 2013. Available at
```
http://th-www.if.uj.edu.pl/~sitarz/Sitarz_Zakopane-1.pdf
```
23. Connes, A. : *Noncommutative Geometry and Physics*. No year. Available at
```
https://www.impan.pl/swiat-matematyki/
notatki-z-wyklado~/connes_ncgp.pdf
```

5 MORE ON ALGEBRAS

5.1 Introduction

This chapter consists mainly of mathematical intermezzos for use in other chapters, brought together here for the sake of their thematic connection.

In quantum theory the obvious algebras, familiar to every mathematical physicist, are algebras of operators in Hilbert space, i.e. C^*-algebras and von Neumann algebras. In classical physics this is less so. For instance, the algebra of smooth functions on the phase space of classical mechanics is not a C^*- or von Neumann algebra.

Even in the quantum world there are situations, in which other, more general algebras play an essential role; in some cases the standard (positive definite) Hilbert space, so far the mathematical background of all quantum theoretical discussions, has to be replaced by an indefinite inner product space, as will be shown in Chapter 10.

5.2 General notions of algebras

For a definition and basic properties of algebras, see the mathematical intermezzo *algebra* in Chapter 2, Section 2.

Remark: The use of complex numbers is essential in quantum theory; classical physics is for the main part expressed by real numbers. A real algebra $\mathcal{A}$ can be complexified to a complex $*$-algebra.

- *Mathematical intermezzo*

Complexification of vector spaces and algebras. Let $\mathcal{V}^{\mathbb{R}}$ be a real vector space. Define the direct sum of two copies of $\mathcal{V}^{\mathbb{R}}$ as

$$\mathcal{V}^{\mathbb{C}} = \mathcal{V}^{\mathbb{R}} \oplus \mathcal{V}^{\mathbb{R}},$$

with as elements pairs (x, y) from $\mathcal{V}$, with obvious addition and scalar multiplication. In this vector space one defines a linear map J as $J(a_1, a_2) = (-a_2, a_1)$.

Problem. Show that $J^2 = -1$, implying that J can be used as an imaginary unit, making $\mathcal{V}^{\mathbb{C}}$ into a complex vector space.

Starting from a real algebra $\mathcal{A}$ one constructs in a similar way a complex algebra $\mathcal{A}^{\mathbb{C}}$, first as a complex vector space and then as a complex algebra by defining multiplication by

$$(a_1, a_2)(b_1, b_2) = (a_1 b_1 - a_2 b_2, a_1 b_2 + a_2 b_1),$$

and finally a $*$-operation by $(a_1, a_2)^* = (a_1, -a_2)$. The unit element is $(1, 1)$.

Problem. Show that all this makes $\mathcal{A}^{\mathbb{C}}$ into an associative complex $*$-algebra with unit, the complexification of the given algebra $\mathcal{A}^{\mathbb{R}}$.

In turn a complex *commutative* $*$-algebra has a real subalgebra – of which it is the complexification – consisting of the elements a with $a^* = a$. Both cases carry the same amount of information. However, the real part of a complex *non-commutative* $*$-algebra is *not* a subalgebra.

- *End of mathematical intermezzo*

5.3. C^*-algebras

C^*-algebras are, together with von Neumann algebras (next section), the most important algebras in quantum theory. They appear as 'abstract algebras', or as 'concrete operator algebras'. According to the important theorem by Gelfand and Naimark, mentioned below, every 'abstract algebra is isomorphic to an algebra of operators in a Hilbert space. For a proper definition we need the notion of a norm.

- *Mathematical intermezzo*

Norms in vector spaces and in algebras. A *norm* in a vector space V is a map $x \mapsto ||x||$ such that

$$||\lambda x|| = |\lambda| \, ||x||, \quad \forall \lambda \in \mathbb{C}, \; x \in V,$$

$$||x + y|| \leq ||x|| + ||y||, \quad \forall x, y \in V \quad \text{(triangle inequality)}$$

$$||x|| \geq 0, \; ||x|| = 0 \; \Rightarrow \; x = 0.$$

A norm gives a topology, defining a notion of convergence. If the last property is absent one has a *seminorm*, usually denoted as $p(\cdot)$ instead of $|| \cdot ||$. In this case one needs an infinite system of seminorms to define a topology. A vector space with such a system is called a *locally convex vector space*. These will be discussed in the mathematical intermezzo "Locally convex spaces and algebras" - Section 5 of this chapter.

A norm in an algebra $\mathcal{A}$ should have the additional property

$$||ab|| \leq ||a|| \, ||b||, \quad \forall a, b \in \mathcal{A}.$$

A normed algebra is called *complete* if and only if every Cauchy sequence is convergent. (A sequence $x_1, x_2, \ldots$ is called a Cauchy sequence whenever $\lim_{m,n} ||x_m - x_n|| = 0$).

- End mathematical intermezzo

- Mathematical intermezzo.

C^*-algebras. A complete normed algebra is a C^*-algebra if and only if its norm has the property

$$||a^*a|| = ||a||^2, \ \forall a \in \mathcal{A}.$$

This innocently looking but very strong relation makes C^*-algebras into one of the most powerful and most widely studied objects in functional analysis. The basis of C^*-algebra theory was laid by two Russian mathematicians, Israel Gelfand and Mark Naimark. Their most important paper [1], mentioned above, treats the theory of representations of C^*-algebras in Hilbert space. The basic modern reference for C^*algebras is the book by Jacques Dixmier [2], a clearly written but rather severe Bourbaki style book. More friendly is the book by William Arveson [3]. There are excellent sets of introductory lecture notes [4] and [5], and a more comprehensive set of notes [6] (all three available on internet).

Projections are important when measuring observables. Many C^*-algebras have too few projections. It is for this reason that the elements of a C^*-algebra in general appear as *pre-observables*.

Problem. Show that the algebra $C([0,1])$, the commutative C^*-algebra of all continuous functions on the real interval $[0,1]$, has only the two trivial projections 0 and 1.

- End mathematical intermezzo

The next type of algebra is better in this respect.

5.3   Von Neumann algebras

- Mathematical intermezzo

- Mathematical intermezzo

Von Neumann algebras. C^*-algebras are 'abstract' algebras; von Neumann algebras are algebras of operators in a Hilbert space, i.e. subalgebras of $B(\mathcal{H})$, the algebra of all bounded operators in a Hilbert space $\mathcal{H}$. The study of these algebras was initiated by John von Neumann in 1921, who published, together with Francis Murray, a long series of fundamental papers on the subject, starting in 1936. See [7]. This work is one of the great pieces of mathematics of the 20th century.

Dixmier wrote also a standard treatise on what he first called W^*-algebra, French edition 1969, English translation 1981 [8], in the same severe style as his book on C^*-algebras. More friendly books are those by Jacob Schwartz [9] and Bruce Blackadar [10] (this complete on internet). There is an excellent introductory set of lecture notes by Vaughan Jones [11] (on internet). A fundamental book on operator algebras is that of Richard

Kadison and John Robert Ringrose [12]; good sets of lecture notes on this subject are [13] and [14] (both on internet), and finally an extensive set of notes that discuss it in the general background of functional analysis [15] (on internet).

Let $\mathcal{H}$ be a Hilbert space and $\mathcal{A}$ an $*$-algebra of bounded operators in $\mathcal{H}$. Denote the collection of all bounded operators in $\mathcal{H}$ that commute with all elements of $\mathcal{A}$ as $\mathcal{A}'$. It is called the *commutant* of $\mathcal{A}$. It is again a $*$-algebra of operators in $\mathcal{H}$. The commutant of $\mathcal{A}'$ is called the *bicommutant* of $\mathcal{A}$, and is denoted as $(\mathcal{A}')'$.

Definition: *A $*$-algebra of operators in $\mathcal{H}$ is a von Neumann algebra if and only if $(\mathcal{A}')' = \mathcal{A}$.*

Note that von Neumann algebras are sometimes called W^*-algebras.

The main advantage of von Neumann algebras over C^*-algebras is that with every selfadjoint element it contains also all the projections which are in this element.

The book by Sakai [16] is an excellent book on both von Neumann algebras and C^*-algebras (complete on internet).

- End of mathematical intermezzo

5.4 General locally convex algebras. Fréchet and LF algebras

Quantum theory and the theory of operators in Hilbert space were developed together; for a mathematical physicist their relation was natural and intimate.

However many algebras in physics, in particularly in classical physics, are neither C^*- nor von Neumann algebras. The best example is the commutative algebra $C_{\mathbb{R}}^{\infty}(\mathcal{M})$, the algebra of smooth real-valued functions on the phase space $\mathcal{M}$ of a classical mechanical system, or rather its complexification $C_{\mathbb{C}}^{\infty}(\mathcal{M})$. For this we need general locally convex algebras, with as first concrete example Fréchet spaces and then LF-spaces.

This goes even further. For a rigorous and satisfactory description of the quantum Maxwell field, even in the free case, one has to embed the Hilbert space of physical states into a more general space – a space with an indefinite inner product; anathema to the average quantum physicist. This will be explained in Chapter 10.

- Mathematical intermezzo

Locally convex spaces and algebras. Hilbert spaces are characterized by an inner product $(\cdot, \cdot)$, which defines a *norm* on vectors as $||\psi|| = \sqrt{(\psi, \psi)}$. A generalization of this is a Banach space, a vector space in which there is no inner product but still a norm. The next generalization is a *locally convex vector space (lc vector space)*, the case where instead of a single norm $||\cdot||$ to define the topology (continuity, convergence, limits), one has an infinite system of *seminorms*, with instead of the condition $||\psi|| = 0 \Leftrightarrow \psi = 0$, the weaker requirement $p_\alpha(\psi) = 0, \forall \alpha \Rightarrow \psi = 0$. For more information on locally convex spaces, see [17], [18] and [19] (all three on internet). For background on the development of locally convex spaces, see the talk by Dieudonné [20] (on internet).

A locally convex algebra (lc algebra) has an underlying lc vector space. Just like for normed algebras the seminorm is multiplicative, i.e. $p_\alpha(ab) \leq p_\alpha(a)\, p_\alpha(b)$. For an lc *-algebra one has $p_\alpha(a^*) = p_\alpha(a)$, for all a in the algebra.

A good reference for general topological algebras is the book by Mallios [21].

One wonders if the possible relation $p_\alpha(a^*a) = (p_\alpha(a))^2$ for an lc *-algebra might lead to strong additional properties, like in the case of a C^*-algebra. This seems not to have been discussed in the literature.

- End of mathematical intermezzo

The most useful locally convex algebras for the particular purpose of this book are *Fréchet algebras*.
- Mathematical intermezzo

Fréchet algebras. A *Fréchet space* is a complete metrizable locally convex space, with a topology defined by a countably infinite family of seminorms $\{p_n(\cdot)\}_n$. For the notion of *metrizable* see the mathematical intermezzo *Topology, topological spaces* - Chapter 5, Section 2. A similar definition holds for Fréchet algebra. A good example of a Fréchet algebra is the algebra of all smooth functions on a closed interval $[a, b]$ of the real line. The notion of Fréchet space (Fréchet algebra) can be generalized to that of *LF-space* (*LF-algebra*), a countable direct sum of Fréchet spaces (Fréchet algebras); a good example of an LF-algebra is the algebra of all smooth functions with compact support on the full real line. (The support of a smooth function is the closure of the smallest set on which the function is different from zero.) LF-spaces need not be metrizable.

A good set of lecture notes about Fréchet spaces is [21]; a second very useful set, which treats Fréchet spaces in the context of general topological vector spaces, is [22] (both on internet). A useful generalization of Fréchet space is an LF-space, a direct sum of countably many Fréchet spaces.

Hilbert space together with its operators remain to most mathematical physicists the natural theatre for quantum theory. In Chapter 10, in which the quantization of the Maxwell field is discussed, it will be argued that this is no longer sufficient. In the last ten years research into the field of general locally convex spaces and algebras has led to much interesting and systematic results. However, application of this to physics has lagged behind, such as to the problem of the Maxwell field.

References

1. Gelfand, I.M., Naimark, M.A.: On the imbedding of normed rings into the ring of operators on a Hilbert space". Math. Sbornik 12, 197-213 (1943).
2. Dixmier, J.: C^*-algebras. North-Holland 1977. (Translated from the original 1969 French editon).
3. Arveson, W.: An Invitation to C^*-Algebra. Springer 1976.
4. NN: The Basics of C^*-Algebras. No year. Available at
   https://www.imsc.res.in/~sunder/psc.pdf
5. Warner, G.: C^*-Algebras. No year. Available at
   http://www.math.washington.edu/~warner/C-star.pdf
6. Renault J.: Lecture notes on C^*-algebras. Lecture notes University of Orléans, France. No year. Available at
   http://www.univ-orleans.fr/mapmo/membres/
   renault/books/IMPA_09.pdf
7. Murray, F.J., von Neumann, J.: On rings of operators. Ann. of Math. 37, 116-229, (1936). This was the first paper in a series, of which later papers appeared in 1937, 1940 and 1947.
8. Dixmier, J.: Von Neumann algebras. North-Holland 1981. Translated from the French edition of 1957. The French edition was the first book on von Neumann algebras. Just as 2 it is a basic reference for the subject. written in the same clear but rather severe Bourbaki style.
9. Schwartz: J.. W*-Algebras. Gordon & Breach 1967.
10. Blackadar, B.: Operator Algebras. Theory of C^*-Algebras and von Neumann Algebras. Springer 2006, 2013. Available as complete book at
    http://wolfweb.unr.edu/homepage/bruceb/Cycr.pdf
11. Jones, V.: Von Neumann Algebras. Lecture Notes 2009. Available at:
    https://math.berkeley.edu/~vfr/MATH20909/
    VonNeumann2009.pdf
12. Kadison, R.V., Ringrose, J.R.: Fundamentals of the Theory of Operator Algebras I. Elementary Theory. Amer. Math. Soc. 1997.
13. Katavolos, A.: Operator algebras: an introduction. Athens university lecture notes 2011. Available at
    http://users.uoa.gr/~akatavol/seminar13/chiosar.pdf
14. Marcoux, L.W.: An Introduction to Operator Algebras. University of Waterloo lecture notes 2005. Available at
    http://www.math.uwaterloo.ca/~lwmarcou/
    Preprints/PMath810Notes.pdf
15. Erdman, J.M.: Functional Analysis and Operator Algebras. An Introduction. Portland State university lecture notes 2015. Available at
    http://web.pdx.edu/~erdman/FAOA/
    functional_analysis_operator_algebras_pdf.pdf
16. Sakai, S.: C^*-Algebrs and W^*-Algebras. Springer 1971. The complete book is available at
    http://alirejali.ir/afiles/up/other/book1/

C-%20algebras%20and%20W-algebras1.pdf

17. NN: Functional analysis – locally convex spaces. Johannes Kepler University Linz lecture notes 2012. Available at
http://www.jku.at/analysis/content/e83492/
e84520/e193753/locconv2_ger.pdf

18. Nagy, G.: Locally Convex Vector Spaces I: Basic Local Theory. Kansas State University lecture notes. 2007-2008. Available at
https://www.math.ksu.edu/~nagy/
func-an-2007-2008/lcvs-1.pdf

19. Garrett, P.: Seminorms and locally convex spaces. University of Minnesota lecture notes 2014. Available at
http://www.math.umn.edu/~garrett/m/fun/
notes_2012-13/07b_seminorms.pdf

20. Dieudonné, J.A.: Recent Developments in the Theory of Locally Convex Vector Spaces. Talk given in Chicago 1953. Available at
http://projecteuclid.org/download/pdf_1/
euclid.bams/1183518281

21. Mallios, A.: Topological Algebras. Selected Topics. North-Holland 1986.

22. Vogt, D.: Lectures on Fréchet spaces. Bergische Universität Wuppertal lecture notes 2000. Available at
http://www2.math.uni-wuppertal.de/
~vogt/vorlesungen/fs.pdf

22. Grubb, G.: University of Copenhagen lecture notes. No year. Available at
http://www.math.ku.dk/~grubb/distb.pdf

6 The GNS Representation

6.1 Introduction

The GNS (Gelfand-Naimark-Segal) representation is a mathematical construction, mainly used in quantum theory. It starts from an extremely simple idea, but leads nevertheless to deep results in physics and mathematics. It is an essential ingredient in the physical interpretation of the theory, given a pair consisting of a *-algebra $\mathcal{A}$ of observables or pre-observables and a state ω (linear functional on this algebra). Its main application is in AQFT (Algebraic Quantum Field Theory), discussed in Chapter 10.

Here we shall give a generalized version, which allows a transparant and elegant description of the algebraic version of Wightman quantum field theory in Chapter 8 and cannot be avoided in the discussion of the Maxwell quantum field in Chapter 9. This was already observed in the Preface and in various other places in this book. The standard version, in which the problems with the Maxwell quantum field cannot be treated, is discussed in [1]. [2] and [3] (all three on internet).

Sufficient for the construction of a GNS representation are the following data: a *-algebra $\mathcal{A}$, complex and with unit element, together with a normalized linear functional ω on $\mathcal{A}$, i.e. with $\omega(1_\mathcal{A}) = 1$.

We distinguish three levels obtained by successively adding data, resulting in stronger properties of the representation.

1. The algebra $\mathcal{A}$ is just a complex *-algebra, with ω a normalized functional on $\mathcal{A}$. This is sufficient to carry out the basics of the GNS construction. The result is a representation of $\mathcal{A}$ by linear operators in a linear space with a non-degenerate inner product. Note that non-degenerateness of an inner product $(\cdot, \cdot)$ on a vector space V means that if for $x \in V$ the inner product $(x, y) = 0$ for all y in V, then x must be zero.

No topological properties are involved on this first level. It is the basic situation, not very useful for applications, but useful for demonstrating the mathematical essentials of the construction.

2. Topological data are added: $\mathcal{A}$ is a topological algebra and the functional ω is supposed to be continuous. The representation space has now a

66

continuous non-degenerate inner product. We earlier called such a space a *pseudo-Hilbert space*. This setup is very useful for understanding the essential mathematical structure of Wightman axiomatic quantum field theory, discussed in Chapters 8, 9 and 10. There is a variety of possible topological algebras, as is discussed in Chapter 5.

3. The algebra $\mathcal{A}$ is a C^*-algebra, and ω is normalized and positive, i.e. $\omega(a^*a) \geq 0$ for every $a \in \mathcal{A}$, which implies continuity. This is the standard setup for AQFT. The fact that it is insufficient to describe fields as the quantum field for the Maxwell theory, has not been realized so far. See for instance [4], a typical paper in this respect. We discuss this in Chapters 9 and 10.

For a catalogue of the various types of relevant algebras, see Chapter 5, in particular the mathematical intermezzo C^*-*algebra* - Chapter 5, Section 2 and *von Neumann algebra* - Chapter 5, Section 3, with the references there.

6.2 The representation

The GNS representation is a typical construction of the sort that mathematicians like: one has given a certain mathematical object, lets it act on itself in a fairly trivial way, obtaining then a new non-trivial and interesting object.

Start with a given $*$-algebra $\mathcal{A}$. Consider a copy of $\mathcal{A}$, to be denoted as $\underline{\mathcal{A}}$, with elements $\underline{a}, \underline{b}, \ldots$. This distinction is useful for keeping a certain conceptual clarity. Define for a fixed element a in $\mathcal{A}$ an action π_0 of $\mathcal{A}$ in $\underline{\mathcal{A}}$ as

$$\pi_0(a)\underline{a_1} = \underline{aa_1}, \quad \forall a_1 \in \mathcal{A}.$$

Note that

$$\pi_0(1_{\mathcal{A}})\underline{a_1} = \underline{a_1}, \quad \pi_0(ab)\underline{a_1} = \underline{aba_1} = \pi_0(a)\pi_0(b)\underline{a_1},$$

so π_0 is indeed a representation of $\mathcal{A}$ by linear operators in $\underline{\mathcal{A}}$.

Representation is a general mathematical notion that plays a role in various theories. It means that a mathematical object, here an algebra, but also a group or a Lie algebra, is realized as a system of linear transformations of a vector space. For a few general remarks on representations of algebras, see the mathematical intermezzo *Algebra* - Chapter 2, Section 2.

So far the representation π_0 does not involve the functional ω. We now use it for defining an inner product $(\cdot, \cdot)^\omega$ on the vector space $\underline{\mathcal{A}}$, according to $(\underline{a}, \underline{b})^\omega = \omega(a^*b)$, in cases 2 and 3 continuous because of the continuity of ω. It is in general degenerate. This means that there may be non-zero elements $\underline{a}$ in $\underline{\mathcal{A}}$ such that $(\underline{b}, \underline{a})^\omega = 0$ for all $\underline{b}$ in $\underline{\mathcal{A}}$. We denote the linear sub space of such vectors as $\underline{\mathcal{N}_0}$. A possible degeneracy can be squeezed out by going to the *quotient space* $\underline{\mathcal{A}}/\underline{\mathcal{N}_0}$.

- Mathematical intermezzo

Quotient space. Let X be a (non-empty) set. A relation between two elements x and y of X is called an *equivalence relation*, with notation $\sim$, if it has the following properties

1. $x \sim x$ (reflexivity),
2. $x \sim y \iff y \sim x$ (symmetry),
3. $x \sim y, \ y \sim z \implies x \sim z$ (transitivity),

Equivalence classes are the subsets of X consisting of all equivalent elements. An equivalence class that contains x is denoted as $[x]$. This means $[x] = [y]$ if and only if $x \sim y$. The set of equivalence classes with respect to $\sim$ is called the *quotient space* over $\sim$ and is denoted as $X/\sim$. There is a rather obvious surjective map

$$\varphi : X \to X/\sim \, , \quad x \mapsto [x].$$

All this can be applied to various mathematical objects. When X is a vector space V, the definition of an equivalence relation is extended:

4. $x \sim y \iff x - z \sim y - z$ (translation invariance)
5. $x \sim 0 \iff \lambda x \sim 0$, for all λ in $\mathbb{R}$ or $\mathbb{C}$,

Problem. Show that the space of equivalence classes is a vector space under addition and scalar multiplication, defined as

$$[x] + [y] := [x + y], \quad \lambda[x] := [\lambda x], \quad \forall x, y \in V, \ \lambda \in \mathbb{C}, \mathbb{R}.$$

The quotient space of the vector space V over a sub vector space V_0 is denoted as V/V_0. The map φ from this space to V_0 is linear. A linear map

$$T : X \to X, \ x \mapsto y$$

descends to a linear map

$$T' : V/V_0 \to V/V_0, \ [x] \mapsto [y]$$

if and only if T leaves V_0 invariant.

Problem. Prove this.

To obtain again an algebra as a quotient space of an algebra $\mathcal{A}$ we need further properties. A *left ideal* in an algebra $\mathcal{A}$ is a sub vector space $\mathcal{I}$ with the property

$$a \in \mathcal{A}, \ b \in \mathcal{I} \implies ab \in \mathcal{I}, \ \forall a \in \mathcal{A}.$$

A *right ideal* in $\mathcal{A}$ is defined as

$$a \in \mathcal{A}, \ b \in \mathcal{I} \implies ba \in \mathcal{I}, \ \forall a \in \mathcal{A}.$$

A *two-sided ideal* is both a left and a right ideal.

The quotient space $\mathcal{A}/\mathcal{I}$ is again an algebra if and only if $\mathcal{I}$ is a two-sided ideal.

Problem. Prove this.

- End of mathematical intermezzo

In the next step in the construction of the GNS representation one lets the operators $\pi(a)$ descend from $\underline{\mathcal{A}}$ to the quotient space $\underline{\mathcal{A}}/\underline{\mathcal{N}_0}$. For this – according to the criterium mentioned above– the operators $\overline{\pi(a)}$ must leave the subspace $\underline{\mathcal{N}_0}$ invariant.

Problem. Show that this is indeed the case.

As a result we have a $*$-representation of the algebra $\mathcal{A}$ in the inner product space $\underline{\mathcal{A}}/\underline{\mathcal{N}_0}$. The inner product in this space is non-degenerate, by construction. We denote this representation space as $\underline{\mathcal{A}}/\underline{\mathcal{N}_0} = \mathcal{H}_0^\omega$.

In case 3 (positive inner product) this space is a pre-Hilbert space, which can then be completed to a Hilbert space $\mathcal{H}^\omega$. This is in fact the standard GNS construction found in the literature. Case 1 is purely algebraic, so there completion makes no sense. In case 2 (indefinite inner product) it will depend on the sort of topological algebra $\mathcal{A}$. It may or may not have a completion.

There is a special vector in the representation space $\mathcal{H}^\omega$; it is the vector $1_{\underline{\mathcal{A}}}$. In the application in quantum field theory it is denoted as Ω_0 and is then called *ground state*, *vacuum state*, or *0-particle state*, with the expression $(\Omega_0, \pi_\omega(a), \Omega_0)$ the vacuum expectation value of the observable a.

The vector $1_{\underline{\mathcal{A}}}$ is *cyclic*, meaning that by applying the operators $\pi(a)$ on it one obtains the full $\mathcal{H}^\omega$.

If the state ω is pure, the GNS representation is irreducible, meaning that there is no non-trivial sub space of the representation space, left invariant by all operators $\pi(a)$.

6.3   Final remark

It does not matter whether the algebra $\mathcal{A}$ in the GNS construction is commutative or non-commutative, i.e. whether it is applied to classical or quantum physics. This is in agreement with the general purpose of this book, which is to give a formalism for a unified description for both classical and quantum physical systems. Nevertheless, the GNS construction has almost exclusively been used for quantum theory. There is however some work in which the GNS construction has been discussed in a classical context [5].

References

1. Jaffe, A.: The GNS construction. Harvard University lecture notes 2015. Available at
   http://isites.harvard.edu/fs/docs/
   icb.topic1466080.files/GNS.pdf
2. NN: Gelfand-Naimark-Segal construction. Lecture notes 2013. Available at
   http://planetmath.org/sites/default/files/
   texpdf/40256.pdf
3. Miyabe, K.: C^*-algebraic methods in spectral theory. Nagoya University lectures. No year. Available at
   http://www.math.nagoya-u.ac.jp/~richard/
   teaching/s2014/Miyabe.pdf
4. Buchholz, D., Ciolli, F., Ruzzi, G., Vasselli, E. : The universal C^*-algebra of the electromagnetic field. 2015. Available at
   http://arxiv.org/abs/1506.06603
5. Pulé, J.V.: A unified approach to classical and quantum K.M.S. theory. Reports on Mathematical Physics. 20 . 1, 75-81 (1984).

7 QUANTUM FIELD THEORY. WIGHTMAN'S APPROACH

7.1 Introduction

Quantum mechanics describes in the first place the physics of atoms and molecules. Its principles were developed in the 1920s by – among others – Werner Heisenberg, Erwin Schrödinger and Max Born. As a physical theory it is completely satisfactory, and so is its mathematical formulation, of which the basis goes back to the work of John von Neumann [1].

In the 1930s atomic physics developed further into nuclear physics, and then in the 1940s and later into the physics of elementary particles, or, as it is usually called, high energy physics.

For this quantum mechanics was no longer sufficient; quantum field theory appeared on the stage.

Quantum mechanics, discussed in Chapter 3, Section 4, is the result of the quantization of classical mechanics, which describes the behavior of systems of (non-relativistic) particles. Quantum field theory comes from the quantization of classical fields, in this context in the first place the electromagnetic field, Maxwell field or radiation field.

There are of course other classical fields, the fields describing hydrodynamics or aerodynamics, and in particular the gravitation field in Einstein's theory of general relativity. At this point it may be noted that the quantization of the gravitational field is still one of the great open problems in present day physics.

Paul Dirac gave in 1927 the first version of a quantized radiation field [2].

7.2 Quantum field theory in particle physics

The way quantum field theory in high energy physics is used is heuristic but very effective. In this formulation the theory then has the same form as ordinary quantum mechanics.

There is a Hilbert state space and field operators, depending on spacetime points x, and describing various types of particles, e.g, $\phi(x)$ (scalar field, π_0-mesons), $\psi(x)$ (spinor field, electrons, protons, neutrons), and $A_\mu(x)$ (vector field, photons), etc..

There is (or there is supposed to be) a Hamiltonan operator, built from these field operators. The theory gives a scattering operator S, which connects the states, i.e. the momenta and energies of the different particles, from $t = -\infty$ to $t = +\infty$. This is all that one wants to know in accelerator experiments. The matrices of this S-operator are calculated by expansion in a power series, with each term represented by a sum of diagrams. These 'Feynman diagrams' stand for multiple integrals over the 4-momenta of the particles that take part in the process. For quantum field theory as it is practiced the basic reference is the three-volume treatise by Weinberg [3] (volume II complete on internet). There is also a short introduction by Weinberg [4] (on internet). Other useful texts are Ryder [5] and Tong [6] (on internet). There are good books by Zee [7] and by Srednicki [8] (complete on internet) and lecture notes by 't Hooft [9] (on internet).

However, most of the integrals in quantum field theory are divergent. This was already noticed in an early stage by – among others – Paul Dirac.

Calculations, approximation by power series, in particular in various processes of what became known as quantum electrodynamics, processes in which photons (particles of radiation) interact with electrons and positrons, the anti-particles associated with electrons, gave infinities, because all the terms were sums of divergent integrals.

A solution, at least for practical purposes, was found in the method of *renormalization*, a way of systematically removing the infinities by equally infinite counter terms. It was developed by Julian Schwinger, Richard Feynman, Shin'ichiro Tomonaga, and Freeman Dyson. The first three received in 1965 the Noble prize for this.

The method was and still is very successful. The experimental results in certain particle processes can be predicted to more than ten decimals. Nevertheless, severe criticism remained and still remains.

Dirac (in 1975): "Most physicists are very satisfied with the situation. They say: "Quantum electrodynamics is a good theory and we do not have to worry about it any more." I must say that I am very dissatisfied with the situation, because this so-called 'good theory' does involve neglecting infinities which appear in its equations, neglecting them in an arbitrary way. This is just not sensible mathematics. Sensible mathematics involves neglecting a quantity when it is small - not neglecting it just because it is infinitely great and you do not want it!"

Feynman, whose role was crucial in the invention of renormalization theory (in 1958): "The shell game that we play ... is technically called 'renormalization'. But no matter how clever the word, it is still what I would call a dippy process! Having to resort to such hocus-pocus has prevented us from proving that the theory of quantum electrodynamics is mathematically self-consistent. It's surprising that the theory still hasn't been proved self-consistent one way or the other by now; I suspect that renormalization is not mathematically legitimate.".

A short review of renormalization can be found at [10] (on internet).

7.3 Wightman's rigorous approach

The first attempt at finding a mathematically rigorous formulation of quantum field theory is due to Arthur Wightman. In his approach a quantum field theory is characterized by all its so-called n-point functions, the vacuum expectation values of the product of n field operators. See his basic paper [11].

The first and still most successful quantum field theory is quantum electrodynamics, the theory of electrons, positrons and photons. It has however a peculiar feature. The requirements of manifest Lorentz covariance and a positive definite inner product state space, i.e. a Hilbert space, cannot be met simultaneously. This is even true for the free Maxwell quantum field, a purely linear theory, that can be analyzed completely. This is the well-known Gupta-Bleuler phenomenon. See for this the references [10], [11], [12] and [13] of Chapter 9. In standard particle physics this is not much of a problem, given the low level of mathematical rigor there. It is however disappointing that Wightman's rigorous approach is not able to cope with it, which means that the Maxwell quantum field is excluded from the Wightman formalism. A generalized form of this formalism in which this case fits will be discussed in Chapter 9.

Note that also the gauge fields, with its basic example Yang-Mills theory, important in present day particle physics, also do not fit in the Wightman scheme, for reasons that are similar to those of the Maxwell field. To exhibit this is however more difficult as there is no linear free field. So far very little has been done with respect to this general problem.

In fact in most books on Wightman's theory, such as, for instance, the book by Streater and Wightman [17], the quantum Maxwell field, the oldest and in its renormalized form the most successful quantum field theory is not even mentioned. We shall observe the same in algebraic quantum field theory in Chapter 10.

For the sake of simplicity we shall restrict the following discussion to a single scalar field $\phi(x)$. The other cases are more complicated but not essentially different. For a scalar field one has a series of functions $W_n(x_1, \ldots, x_n) = (\Omega_0, \phi(x_1) \cdots \phi(x_n)\Omega_0)$, for $n = 0, \ldots$, with Ω_0 the 'vacuum vector', i.e. the vector which describes the 0-particle state, and the x_j points in 4-dimensional spacetime.

At this point it should be noted that, strictly speaking, field operators at sharp points do not exist. In a rigorous formulation one should instead use operators linearly depending on smooth functions, e.g. $\phi(f)$, heuristically written as $\phi(f) = \int_{\mathbb{R}} \phi(x)f(x)dx$, i.e. operator-valued generalized functions.

- Mathematical intermezzo

Generalized functions, Dirac δ-function. The δ-function was invented by Dirac, a genius in very effective heuristic mathematical notions. Another example is his *bra-ket* formalism, equally popular in physics text books.

Dirac defined his δ-function $\delta(x)$ as a function which is zero outside $x = 0$ and infinity in $x = 0$ It is further defined by the integration formula

$$\int_{\mathbb{R}} \delta(x)f(x)dx = f(0),$$

for all smooth functions f ('test functions'). There is a first derivative $\delta^{(1)}(x)$ defined by the relation

$$\int_{\mathbb{R}} \delta^{(1)}(x)f(x)dx = f^{(1)}(0).$$

Higher derivatives are defined in an analogous manner. The δ-function itself is the derivative of the *step function* $H(x)$ which is 0 for $x < 0$ and 1 for $x \geq 0$.

Problem. Derive the more general formulas

$$\int_{\mathbb{R}} \delta(x - x_0)f(x)dx = f(x_0),$$

$$\int_{\mathbb{R}} \delta^{(1)}(x - x_0)f(x)dx = -f^{(1)}(x_0).$$

Problem. Show that the step function is indeed the derivative of the δ-function in the above sense. Explain the minus sign.

The step function and an early version of the δ-function was invented by Oliver Heaviside (1850-1925), an electrical engineer who made important contributions to a wide area of mathematics, physics and technology. He gave a rigorous formulation of the δ-function as a series of functions. Let $\{g_n(x)\}_{n=0,1,2,...}$ be a series, such that

$$\int_{\mathbb{R}} g_n(x - x_0)dx = 1, \quad \forall n = 0, 1, \ldots,$$

and

$$\lim_{n \to \infty} \int_{n \to \infty} g_n(x - x_0)f(x)dx = f(x - x_0),$$

for every fixed point x_0 and all smooth test functions f. This (rigorous) formula gives the same result as Dirac's (heuristic) δ-function formula.

Much later the French mathematician Laurent Schwartz (1915-2002) developed a general theory of generalized functions, or 'distributions', as he called them, within a functional analytic framework, with the idea that distributions are linear functionals on test function spaces.

Excellent lecture notes on the Dirac δ-function are [12], [13] and [14] (all three on internet) and on generalized functions [15] (on internet).

- End of mathematical intermezzo

Note also that the operators $\phi(f)$ are unbounded, i.e. only defined on dense domains in the Hilbert space $\mathcal{H}$. (See for the notion of unbounded operator the mathematical intermezzo *Operators in Hilbert space* - Chapter 3, Section 5). For this reason one has to require a common invariant dense domain for all operators $\phi(f)$.

Another requirement is *locality*, meaning that in a relativistic theory, fields at points x, y that are space-like with respect to each other, i.e. with $(x_0 y_0 - x_1 y_1 - x_2 y_2 - x_3 y_3) \leq 0$ should commute. This should, of course, be translated into a condition in terms of supports of test functions.

In the remainder of this chapter we will, for the sake of convenience and transparency, not use test functions but rather the heuristic language with sharply defined quantities at spacetime points. However, in the next chapter test functions will be essential.

7.4   Properties of Wightman's n-point functions

We restrict the discussion to the simplest case of a single real scalar field $\phi(x)$, enough to understand the general situation. The n-point functions, defined as

$$W_n(x_1, \ldots, x_n) = (\Omega_0, \phi(x_1) \cdots \phi(x_n) \Omega_0)$$

reflect the properties of the theory, in the first place its Lorentz covariance. This means that there is a unitary representation of the Poincaré group acting in the Hilbert space of the theory such that

$$U(a, \Lambda)\phi(x)U(a, \Lambda)^{-1} = \phi(a + \Lambda x),$$

leaving invariant the vacuum vector Ω. This implies that the covariance properties of the Wightman functions under the Poincaré group are

$$W_n(a + \Lambda x_1, \ldots, a + \Lambda x_n) = W_n(x_1, \ldots, x_n).$$

(See for the Poincaré group the Mathematical intermezzo *Poincaré group or inhomogeneous Lorentz group* - Chapter 2, Section 6). There are further properties, that, for instance, express the positivity of the inner product of the Hilbert space and of the energy spectrum and a suitable form of the locality condition, mentioned in the proceeding section.

The Wightman theory as a whole is formulated in a system of axioms, the *Wightman axioms*. Systems of such axioms, for various fields, connected with various elementary particles and their interactions describe all possible relativistic quantum field theories.

Within Wightman's rigorous formulation of field theory some general theorems have been derived, for instance the so-called *CPT - Spin Statistics Theorem*, which connects symmetry under the product of charge conjugation, space reflection and time reversal with the distinction between bosons (particles with integer spin) and fermions (particles with half integer spin).

Nevertheless, Wightman's approach has not been able to solve the general mathematical problems of quantum field theory, with its infinities in the calculated results. Almost the only field theories that are known to fit in his scheme are the free fields.

There are two books on Wightman theory, a very comprehensive book by Bogolubov et al. [16], and an introductory, more accessible book by Streater and Wightman [17].

A next generation of mathematical physicists has tried to come to grips with this situation by constructing simpler models, with interaction, but with space and momentum cut-offs and in lower spacetime dimensions. This is called *constructive quantum field theory (CQFT)*, with as pioneers Arthur Jaffe and James Glimm. This involved, among other techniques, using models with spacetime cut-offs, in which one then tries to establish the existence of the limit for infinite space, and studying models in lower spacetime dimensions. Interesting and meaningful results have been obtained, but nothing in four dimensions. See for an overview of CQFT an article by Jaffe [18] (on internet).

7.5   The reconstruction theorem

A system of Wightman functions reflects the properties of a quantum field, but it does more. Given a system of Wightman functions which satisfies the Wightman axioms, a unique full Hilbert space theory, with the vacuum state, the field operators together with the unitary operators that represent the Poincaré group, can be constructed. This is stated by the *reconstruction theorem*. It is a highlight in Wightman's approach; its proof is complicated and non-trivial.

In the next chapter we shall introduce a generalization of Wightman's formulation which brings it into the range of the theme of this book, namely as a an *algebraic covariance system*. In this the reconstruction theorem becomes natural and easy to prove, with the proof being a generalization of the GNS representation discussed in the preceding chapter.

7.6   CPT invariance

As an attempt to put relativistic quantum field on a mathematically rigorous basis, Wightman axiomatic field theory is very convincing. It is a beautiful theory, even more so in its algebraic reformulation discussed in the next chapter.

Nevertheless, no realistic theory, i.e. one with non-trivial interaction between particles, has been explicitly constructed in terms of Wightman's approach.

There is however an experimentally observed property which has been derived from it: general CPT invariance. This is the product of three possible discrete symmetries: *charge conjugation* (C), *parity*, i.e. the symmetry connected with space reflection (P) and *time reversal* (T). Each of these quantities may separately be conserved for at most one of two of these three. All field theories have the product CPT as a conserved quantity. This follows from the analiticy properties of the Fourier transforms of the Wightman functions and has been verified experimentally. For an overview of discrete symmetries, see [19] (available at internet).

References

1. von Neumann, J. Mathematical Foundations of Quantum Mechanics. Translated from the original 1932 German edition. Princeton 1955.
2. Dirac, P.A.M.: The Quantum Theory of the Emission and Absorption of Radiation. Proc. Royal Soc. (London) **A114**, 243D265, (1927) Available at:
   http://dieumsnh.qfb.umich.mx/
   archivoshistoricosMQ/ModernaHist/
   Dirac1927.pdfDirac1927
3. Weinberg, S.: The Quantum Theory of Fields (3 volumes). Cambridge University Press 2005. The complete Volume II is available at
   http://www.fulviofrisone.com/attachments/article/453/
   Weinberg,%20Steven%20-%20The%20Quantum
   %20Theory%20Of%20Fields%20Volume%20II.pdf
4. Weinberg, S.: What is Quantum Field Theory, and What Did We Think It Is?. 1997. Available at
   http://arxiv.org/pdf/hep-th/9702027.pdf
5. Ryder, L. H.: Quantum Field Theory. Cambridge University Press 1985.
6. Tong, D.: Quantum Field Theory. DAMTP Cambridge lecture 2006-2007. Available at
   http://www.damtp.cam.ac.uk/user/tong/qft/qft.pdf
7. Zee, A.: Quantum Field Theory in a Nutshell. Princeton University Press 2003.
8. Srednicki, M.: Quantum Field Theory. Cambridge 2007. Available at
   http://web.physics.ucsb.edu/~mark/ms-qft-DRAFT.pdf
9. 't Hooft, G.: The Conceptual Basis of Quantum Field Theory. Utrecht University lecture notes 2004. Available at
   http://www.staff.science.uu.nl/~hooft101/
   lectures/basisqft.pdf
10. NN: Renormalization of interacting quantum field theories. University of Leipzig. 2017. Available at
    http://home.uni-leipzig.de/tet/?page_id=283
11. Wightman, A.S.: Quantum Field Theory in Terms of Vacuum Expectation Values. Phys. Rev. 101, 860-866 (1956).
12. NN: The Dirac Delta Function, $\delta(x - x_0)$. Lecture notes. Available at
    http://web.mst.edu/~hale/courses/411/411_notes/

78

`Chapter1.Appendix.Dirac.Delta.pdf`
13. Dray, T.: The Dirac Delta function. Oregon State University Lecture notes. No year. Available at
`http://physics.oregonstate.edu/~tevian/`
`paradigm0/delta.pdf`
14. NN: The Dirac Delta Function. Lecture notes. No year. Available at
`http://www.nada.kth.se/~annak/diracdelta.pdf`
15. Wilde, I.F. : Distribution Theory (Generalized Functions). No year. Available at
`http://homepage.ntlworld.com/ivan.wilde/`
`notes/gf/gf.pdf`
16. Bogolubov, N.N., Logunov, A.A., Oksak, A.I., Todorov, I.: General Principles of Quantum Field Theory. Springer 1990.
17. Streater, R.F., Wightman, A.S.: PCT, Spin and Statistics, and All That. Princeton 2000.
18. Jaffe, A.: Constructive Quantum Field Theory. Harvard University preprint. No year. Available at
`http://www.arthurjaffe.com/Assets/pdf/CQFT.pdf`
19. Van den Brand, J.: Discrete Symmetries. Lectures Vrije Universiteit Amsterdam. Available at
`http://www.nikhef.nl/~jo/CP_course.pdf`

8 ALGEBRAIC WIGHTMAN THEORY

8.1 Introduction

In this chapter we shall present an algebraic reformulation and in the next chapter an algebraic generalization of Wightman quantum field theory by which it becomes an *algebraic covariance system*. Because of this it falls in line with the general theme of this book. The idea for this algebraization is due to Hans-Jürgen Borchers and Armin Uhlmann, published probably independently in 1962 [1] and [2] (on internet, in German). See also a later, much clearer and more explicit paper by Wolfgang Lassner and Armin Uhlmann [3] (on internet) and a clear summer school exposition by Walter Wyss [4]. In our opinion this *Borchers-Uhlmann formalism* definitely deserves more attention than it has gotten so far.

8.2 The algebra of observables

An important part of the mathematics needed for this section will be sketched in the following intermezzo.

- Mathematical intermezzo

Tensor product. Tensor algebra. For every pair of vector spaces V and W one can construct a new vector space, called the *tensor product of V and W* and denoted as $V \otimes W$.

For finite dimensional vector spaces this tensor product can be constructed in the following way. Choose bases $\psi_1, \ldots, \psi_m$ in V and $\phi_1, \ldots, \phi_n$ in W. Consider $m \times n$ objects $\psi_j \otimes \phi_k$, for $j = 1, \ldots, m$ and $k = 1, \ldots n$. The collection of all formal linear combinations of these objects form in an obvious manner an $(m \times n)$-dimensional vector space. This is $V \otimes W$.

Problem. Show that $V \otimes W$, thus defined, is in fact independent of the choice of the bases $\{\psi_j\}_j$ and $\{\phi_k\}_k$.

Hint. Consider two basis transformations, one from $\{\psi_j\}_j$ to a new basis $\{\psi'_j\}_j$ and the other from $\{\phi_k\}_k$ to a new basis $\{\phi'_k\}_k$. These transformations can be used to define new objects $\psi'_j \otimes \phi'_k$, with, by taking linear combinations, gives also a vector space. Show that there is a unique linear isomorphism between these two spaces, which proves that the definition of $V \otimes W$ is coordinate independent.

The definition of $V \otimes W$ for arbitrary (infinite-dimensional) vector spaces is more subtle. There are separate proofs for its existence and for its uniqueness.

Along the lines indicated above one defines $V_1 \otimes \cdots \otimes V_n$, the tensor product of an arbitrary (finite) number of vector spaces. The tensor product is *associative*, for instance,

$$(V_1 \otimes V_2) \otimes V_3 = V_1 \otimes (V_2 \otimes V_3) = V_1 \otimes V_2 \otimes V_3.$$

The n-fold tensor product $V \otimes \cdots \otimes V$ is usually written as an nth tensor power $\otimes^n V$.

The construction of the tensor product $\mathcal{H}_1 \otimes \mathcal{H}_2$ of two Hilbert spaces $\mathcal{H}_1$ and $\mathcal{H}_2$ leads in first instance to a pre-Hilbert space, which, by adding the limits of Cauchy sequences, can be completed to a Hilbert space, sometimes denoted as $\mathcal{H}_1 \widehat{\otimes} \mathcal{H}_2$. In order not to burden the notation we shall denote this completed tensor just as $\mathcal{H}_1 \otimes \mathcal{H}_2$.

Let V be a vector space. The infinite direct sum

$$T(V) = \oplus_{n=0}^{\infty} (\otimes^n V),$$

is an associative algebra, the *tensor algebra over* V. In this sum V^0 is a 1-dimensional vector space, with a fixed unit vector Ω_0. Elements in $T(V)$ are arbitrary linear combinations of elements of the form $f_1 \otimes \cdots \otimes f_k$.

For more details on tensor products see [5], [6] and [7] (all three on internet), and, more particular, for tensor product of Hilbert spaces [8] (on internet).

- End of mathematical intermezzo

With the material from this intermezzo we can discuss the algebras that we need for the algebraic formulation of Wightman quantum field theory, or as we shall call it, the *Borchers-Uhlmann formalism*.

The algebra of observables $\mathcal{A}$, or rather of pre-observables, of a real scalar Wightman quantum field theory in the algebraic formalism is the tensor algebra $T(V)$ with V the space of complex-valued Schwartz scalar test functions on spacetime.

A Schwartz test function is an infinitely differentiable function which goes, together with all its derivatives, to zero, faster than any inverse polynomial. Such functions have the advantage that their momentum space Fourier transforms are also Schwartz functions.

Note that the algebra $\mathcal{A} = T(V)$ is in an obvious way a $*$-algebra.

This is the simplest case. A field theory with various other fields, such as, for instance, vector and spinor fields, is built on tensor algebras of spaces of multi-component test functions.

8.3   States

A state of a Wightman quantum field in the algebraic formalism is a linear functional $\omega : \mathcal{A} \to \mathbb{C}$. *It is*
  - *real, i.e. with* $\omega(a^*) = \overline{\omega(a)}$, *for all* $a \in \mathcal{A}$
  - *positive, i.e.* $\omega(a^*a) \geq 0$, *for all* $a \in \mathcal{A}$,
  - *normalized, i.e.* $\omega(1_\mathcal{A}) = 1$.

Using positivity, it can easily be shown that a state is continuous.

8.4   Interpretation

We should realize that the field operators $\phi(f)$, although the basic observables, theoretically speaking, are not observables that can be measured experimentally. Fields are not observables. The physical observables that are measured in the scattering experiments in particle accelerators, employed in high energy physics, are energy and momentum of incoming and outgoing particles and particle numbers. These are fairly complicated expressions in the fields, in which we do not enter here. The interpretation of a state together with such an observable follows the usual procedure of quantum theory.

For the physical interpretation of a pair (a, ω) we could use the generalized version of the GNS (Gelfand-Naimark-Segal) representation, discussed in Chapter 7, Section 2 (level 2). This would give a Hilbert space $\mathcal{H}^\omega$, with the vacuum / 0-particle state Ω_0 as a vector state. It is however not very useful.

8.5   Symmetries

The main symmetry is connected with the Poincaré group.

It acts by $*$-automorphisms of the algebra of fields as

$$U(\Lambda, a)\phi(f)U(\Lambda, a)^{-1} = \phi((\Lambda, a)f),$$

with the multiplication rule $(\Lambda_2, a_2)(\Lambda_1, a_1) = (\Lambda_2\Lambda_1, \Lambda_2 a_1 + a_2)$.

Problem. Show that the alternative formula

$$U(\Lambda, a)^{-1}\phi(f)U(\Lambda, a) = \phi((\Lambda, a)f),$$

is inconsistent with the multiplication formula.

Time development is included in Lorentz symmetry, because the splitting up of 4-dimensional spacetime is coordinate dependent. Relativistic quantum field theory is therefore, strictly speaking, not an algebraic dynamical system, but an algebraic covariance system.

References

1. Borchers, H.-J.: On Structure of the Algebra of Field Operators. Nuovo Cimento XXIV, 214-236 (1962).
2. Uhlmann, A.: Über die Definition der Quantenfelder nach Wightman und Haag. Wiss. Zeitschrift der Karl-Marx-Universität Leipzig 11, 213-217 (1962). Available (in German) at
   http://www.physik.uni-leipzig.de/~uhlmann/
   PDF/Uh62a.pdf
3. Lassner, G., Uhlmann, A.: On Positive Functionals on Algebras of Test Functions for Quantum Fields. Commun. math. Phys. 7, 152-159 (1968). Available at
   https://projecteuclid.org/download/pdf_1/
   euclid.cmp/1103840378
4. Wyss, W.: On Wightman's Theory of Quantized Fields. In "Boulder Lectures in Theoretical Physics". Gordon and Breach 1969.
5. Hitchin, N.: Tensor Products. Oxford University Lecture Notes. Available at
   https://people.maths.ox.ac.uk/hitchin/hitchinnotes/
   Differentiable_manifolds/Chapter_2.pdf
   These notes are part of a full course on differential manifolds. 2012. available at
   http://people.maths.ox.ac.uk/hitchin/hitchinnotes/
   Differentiable_manifolds/manifolds2012.pdf
6. Ganatra, S. : Tensor Products. Lecture Notes, Stanford University 2013. Available at
   http://math.stanford.edu/~ganatra/math113/
   notes/tensor_products.pdf
7. Hackl, K., Goodarzi, M.: A Small Compendium on Vector and Tensor Algebra and Calculus. Ruhr University Bochum Lectures notes 2010. Available at
   http://www.vgu.edu.vn/fileadmin/pictures/
   studies/master/compeng/study_subjects/modules/
   cm/lecture_notes_-_math.pdf
8. Berberian : Tensor product of Hilbert spaces. University of Texas Lecture Notes. Available at
   https://www.ma.utexas.edu/mp_arc/c/14/14-2.pdf

9
Wightman Theory for the Maxwell Field

9.1 Introduction. Classical Maxwell theory

In standard textbooks classical electromagnetism in vacuum is usually described by two 3-dimensional vector fields, the electric field $\mathbf{E}(\mathbf{x}, t)$ and the magnetic field $\mathbf{B}(\mathbf{x}, t)$. There exist moreover a 3-dimensional vector potential $\mathbf{A}(\mathbf{x}, t)$ and a scalar potential $\phi(\mathbf{x}, t)$ from which the two vector fields $\mathbf{E}(\mathbf{x}, t)$ and $\mathbf{B}(\mathbf{x}, t)$ can be derived. They are auxiliary fields, without a direct physical meaning, but nevertheless useful, in fact indispensable, for deriving Maxwell's equations by use of the Lagrange formalism. We shall not discuss this. The potentials were, of course, unknown to Maxwell.

One of the main arguments for Einstein in his development of the special theory of relativity, as put forward in his 1905 paper, was that the general theory of electromagnetism, due to Maxwell, is already a relativistic theory, even though this is not immediately clear in the standard 3-dimensional formulation.

The best way to see this essential relativistic character of electromagnetism is to write Maxwell's equation in 4-dimensional spacetime form. In this formulation there is a single anti-symmetric covariant 2-tensor field $F_{\mu\nu}(x)$. Maxwell's equations take the form

$$\partial_\mu F_{\nu\rho}(x) + \partial_\nu F_{\rho\mu}(x) + \partial_\rho F_{\mu\nu}(x) = 0, \qquad (9.1.1)$$

with, for the field in vacuum, the dynamical equation

$$\partial_\mu F^{\mu\nu}(x) = 0. \qquad (9.1.2)$$

The potentials $\mathbf{A}(\mathbf{x}, t)$ and $\phi(\mathbf{x}, t)$ together form one 4-dimensional vector field $A_\mu(\mathbf{x}, t)$. The six components of the field tensor $F_{\mu\nu}(\mathbf{x}, t)$ are related to the three components of the electric field $\mathbf{E}(\mathbf{x}, t)$, together with three components $\mathbf{B}(\mathbf{x}, t)$ of the magnetic field. The relation between the tensor field and the vector field, which we shall call the *basic relation*, is

$$F_{\mu\nu} = \partial_\mu A_\nu - \partial_\nu A_\mu. \qquad (9.1.3)$$

Before discussing the further properties of these equations, we should explain the 4-dimensional notation and conventions of relativity theory used in the remainder of this chapter.

- *Mathematical intermezzo*

Relativistic conventions and notation. Spacetime points are denoted as x, meaning $x = (x^0 = ct, \mathbf{x} = x^1, x^2, x^3)$. Partial differentiation $\partial_\mu = \frac{\partial}{\partial x^\mu}$. There is a metric tensor $g_{\mu\nu}$ with $g_{00} = 1, g_{11} = g_{22} = g_{33} = -1$, and with the other elements 0. The metric tensor has this trivial form only in the special theory of relativity; in general relativity where spacetime is no longer a flat 4-dimensional linear space, but a manifold with curvature, it depends on x. The metric tensor is used to connect covariant (lower indices) with contravariant ones (upper indices). There is also contravariant metric tensor which has the same elements as the given covariant one. Again, this is only so in special relativity.

For example, for vector fields $A_\mu(x)$ (covariant) and $A^\mu(x)$ (contravariant) one has

$$A^\mu(x) = g^{\mu\nu} A_\nu(x), \qquad A_\mu(x) = g_{\mu\nu} A^\nu(x).$$

In these formulas the *Einstein summation convention* is used, i.e. the $\sum$-sign is omitted when summing over pairs of the same covariant and contravariant indices.

Problem. Prove the relations

$$g_{\mu\nu} g^{\rho\sigma} = \delta^\rho_\mu \delta^\sigma_\nu, \qquad g^{\mu\nu} = g^{\mu\rho} g^{\nu\sigma} g_{\rho\sigma}.$$

These relations hold also in the less trivial situation of general relativity.

See for information on the Poincaré and Lorentz groups, the basic groups for the special theory of relativity the mathematical intermezzo - Poincaré or inhomogeneous Lorentz group, to be found in Chapter 2, Section 6. For a useful overview of relativistic notation see [1] (on internet).

- *End of mathematical intermezzo*

9.2 Gauge transformations. The Lorenz condition

Different vector fields $A_\mu(x)$ can be associated with the same tensor field $F_{\mu\nu}(x)$. They are connected by *gauge transformations*, i.e.

$$A'_\mu(x) = A_\mu(x) + \partial_\mu \phi(x), \qquad (9.2.1.)$$

with $\phi(x)$ an arbitrary scalar function. A choice of A_μ is called a *gauge*. The collection of gauges for a given tensor field can be restricted by requiring the *Lorenz condition*.

$$\partial^\mu A_\mu(x) = 0. \qquad (9.2.2.)$$

(Note "Lorenz", i.e, Ludvig Lorenz (1829-1891), not "Lorentz", i.e. Hendrik Antoon Lorentz (1853-1928), the Dutch theoretical physicist, who was and still is much more widely known. In the literature confusion on this abounds).

The Lorenz condition also restricts the scalar functions $\phi(x)$ appearing in the gauge transformations. They have to satisfy

$$\partial_\mu \partial^\mu \phi(x) = 0, \qquad (9.2.3.)$$

the zero mass Klein-Gordon equation.

Equation (9.1.1) does not only follow from (9.1.3), but the two equations are equivalent. To understand and derive the above equations, in particular the equivalence (9.1.1) $\longleftrightarrow$ (9.1.3), it is useful to explain the differential geometric background of classical Maxwell theory. For this we need *differential forms*, a subject of great mathematical beauty, worth knowing on its own.

- Mathematical intermezzo

Differential forms. See first, for background, the mathematical intermezzo - Differential manifolds, differential geometry, to be found in Chapter 5, Section 4, together with the references in Chapter 5.

Let $\mathcal{M}$ be an n-dimensional differential manifold. On this there are *differential forms*, i.e. antisymmetric covariant tensor fields

$$B^0(x),\ B^1_{\mu_1}(x),\ B^2_{\mu_1 \mu_2}(x),\ \ldots,\ B^{n-1}_{\mu_1 \mu_2 \cdots \mu_{n-1}}(x).$$

Note that $B^0(x)$ is a scalar function, so has a single component, and that $B^{n-1}(x)_{\mu_1 \cdots \mu_{n-1}}$ has n dependent components, with the result that it is also characterized by only one function.

There is a finite direct sum of the spaces of differential forms, called the *cohomology* of $\mathcal{M}$,

$$\Omega(\mathcal{M}) = \Omega^0(\mathcal{M}) \oplus \Omega^1(\mathcal{M}) \oplus \ldots \oplus \Omega^n(\mathcal{M}),$$

with a product $\wedge$ which makes $\Omega(\mathcal{M})$ into an anti-symmetric tensor algebra. See the mathematical intermezzo - *"Tensor products and tensor algebras"* - Chapter 9, Section 2.

Cohomology theory is a broad and important general field in modern mathematics, in which we cannot enter here. See however [2] (on internet). The particular form used for the descriptions and study of differential forms on a manifold is called the *de Rham cohomology*. See [3] (on internet).

The total cohomology space is a direct sum

$$B = \oplus_{j=0}^n B_j.$$

The separate spaces of forms B^n are mapped into each other by maps $d^p : B^p \to B^{p+1}$, which together form a linear map $d : \Omega(\mathcal{M}) \to \Omega(\mathcal{M})$, the *exterior derivative*, with the properties

$d^2 = 0$ (*idempotency*),

$d(A^p \wedge B^q) = (dA^p)B^q + (-1)^p A^p \wedge (dB^q)$ (*Leibniz property*).

A form B with $dB = 0$ is called *closed*. A form B that is the exterior derivative of a second form A, i.e. $B = dA$, is called *exact*. From $d^2 = 0$ it follows immediately that an exact form is closed.

The quotient space Ker Ω^{p+1}/Ω^p is called the $(p+1)^{th}$ *cohomology space* of $\mathcal{M}$. The cohomology spaces of the manifold $\mathcal{M}$ describe its geometric characteristics. For a reminder of the notion of quotient space, see the mathematical intermezzo - Quotient spaces, to be found in Chapter 7, Section 3.

For the geometrically trivial case of $\mathcal{M} = \mathbb{R}^n$ all cohomology spaces are trivial, i.e. equal to $\{0\}$. In this case there is an equivalence: A closed $\Longleftrightarrow$ A exact.

Differential forms are discussed in all books on differential geometry. See the references of Chapter 5. An excellent book on differential forms is [4]. Good texts on differential forms and the exterior derivative are [5], [6] and [7] (all three on internet). A book on exterior differential system is that by Bryant et al. [8] (complete on internet).

End of mathematical intermezzo

9.3 The free classical Maxwell field in terms of forms

We now formulate the classical theory of the free Maxwell field in the language of differential forms.

1. The tensor field $F_{\mu\nu}$ is a 2-form F. Equation (9.1.1.) becomes in this language $dF = 0$, i.e. F is closed.

2. Spacetime as a manifold is trivial, i.e. just $\mathbb{R}^4$, so a closed form is exact. There is therefore a 1-form A such that $F = dA$. This is equation (9.1.3). So (9.1.1) and (9.1.3) are equivalent.

3. The kernel (null space) of d^0 consists of 0-forms, i.e. scalar functions ϕ. This explains formula (9.2.1).

4. For the dynamical equation (9.1.2) there is also a formulation in terms of differential forms. For this we need an additional notion, *Hodge duality*. This will not be discussed here. See however [9] (on internet).

9.4 The free Maxwell quantum field

Free quantum fields are linear systems, and as such can be analyzed in all detail. Free field theories are not interesting physically; they serve as point of departure for setting up the more interesting cases of fields with interactions. They do in any case not have the divergence problems of interacting field theories. In the language of Wightman theory they are completely determined by their 2-point function, i.e. in heuristic language $W(x_1, x_2) = (\Omega_0, \phi(x_1)\phi(x_2)\Omega_0)$ for the real scalar field. Their state space is a Fock many-particle space, a symmetric or anti-symmetric tensor algebra over a 1-particle Hilbert space. This is also true for the free Maxwell field. This is however meaningless for the gauge quantum fields, such as the Yang-Mills field, which have become very important in recent years. So far a proper Wightman formulation for such fields is absent from the literature.

The Maxwell quantum field has a peculiar feature. The standard requirement in relativistic quantum field theory, i.e. manifest Lorentz covariance, cannot be met unless Hilbert space as state space for the A_μ field is dropped and replaced by some kind of indefinite inner product space. This has be known for a long time, even though it is hard to find a rigorous proof for it. It was first studied independently in 1950, by Suraj Gupta [10] and Konrad Bleuler [11] (on internet). The 1974 paper by Franco Strocchi and Arthur Wightman [12] discussed this problem as a side topic. Their use of a so-called Krein space, a space with both a positive and an indefinite inner product, is inelegant and unsatisfactory; the unitary operators that represent the Poincaré group are unbounded with respect to the Krein Hilbert space structure.

In two papers [13] the author of the present book suggested to use the indefinite inner product space as the correct state space for the A_μ field, without an additional Krein space structure, using the argument that A_μ is an auxiliary field without a direct physical meaning, even though the vector field is necessary for the formulation of the dynamics. He considered two state spaces, one space $\mathcal{H}_A$ with indefinite inner product, in which both $A_\mu(x)$ and $F_{\mu\nu}(x)$ act, and a second space $\mathcal{H}_F^{(0)}$, the quotient space $\mathcal{H}_A/\mathcal{H}_F^{(0)}$, a pre-Hilbert space which can be completed to the physical Hilbert space $\mathcal{H}_F$. The fields $A_\mu(x)$ and $F_{\mu\nu}(x)$ act in $\mathcal{H}_A$; only the physical field $F_{\mu\nu}(x)$ descends to $\mathcal{H}_F^{(0)}$ and then acts by extension in $\mathcal{H}_F$. These two papers have not had any follow up in the literature. If the problem of the Maxwell quantum field is mentioned at all, it is [12] that is the standard reference.

There are two equivalent ways of proving rigorously the properties of the quantum Maxwell field, just mentioned.

1. Construction of a 1-particle space for the vector field operator, in which the Poincaré group acts.
2. Find the Poincaré invariant 2-point function for the vector field operator. It will be unique up to a normalization factor.

Proofs of this, which are complicated, but not essentially deep or difficult, will be found in a forthcoming paper [14].

We will give sketch of method 1 which is to preferred because it is constructive.

Consider first the vector potential $A_\mu(x)$. Following the ideas of the Wightman formalism in the Borchers-Uhlmann approach (See the preceding chapter), we define first the 'abstract' field algebra for the A-field, which we denote as $\mathcal{A}^A$. It is nothing but the tensor algebra generated by the 4-component Schwartz test functions $f^\mu(x)$, or, in other words, by algebra elements $\varphi(f)$, with f 4-components Schwartz functions. This is a locally convex $*$-algebra. Next we consider a state ω^A, a real, continuous, normalized, but not necessary positive functional on $\mathcal{A}^A$. We use the GNS prescription (Chapter 7) to construct a representation space $\mathcal{H}_\omega^A$, a space with a continuous, non-degenerate but not necessarily positive inner product $(\cdot,\cdot)_\omega^A$. It is a "pseudo-Hilbert space" in our terminology. This means that we have a state space in which the potential field operators act, together with the tensor field operators which can be derived from the potential field operators by formula 10.1.3 in this chapter. The Poincaré group acts in an obvious way on the test functions f^μ, and induce in this way $*$-automorphisms $\phi_{(a,\Lambda)}$ on $\mathcal{A}^A$. Manifest Poincaré covariance requires that the state ω^A is invariant, i.e. that $\omega(\phi_{(a,\Lambda)}b) = \omega(b)$, for all b in $\mathcal{A}^A$. This implies that the Poincaré automorphisms are (pseudo-)unitarily implementable, which means that there is a 1-parameter system of (pseudo-)unitary operators $\{U_\omega(t)\}_{t\in\mathbb{R}}$ $U(a,\Lambda)$ acting in the GNS representation space $\mathcal{H}_\omega^A$, such that

$$\pi(\phi_t)(a) = U(t)\pi(a)(U(t))^{-1}, /, \forall t \in \mathbb{R}, \ \forall a \in \mathcal{A}^A.$$

This finishes the discussion of the vector potential.

The impossibility of combining manifest Poincaré covariance with a state space with a positive definite inner product, has far reaching consequences for the formulation of algebraic quantum field theory. This will be discussed in Chapter 10.

References

1. Hagedorn, R.: Introduction to field theory and dispersion relations. Fortschrit. Phys. 5 suppl. 1-127 (1963). Appendix 9. Relativistic notation. Available at
   http://inspirehep.net/record/19314/files/
   Appendix9.pdf
2. Tillmann, U.: Cohomology Theories. Lecture notes Oxford University. No year. Available at
   https://people.maths.ox.ac.uk/tillmann/coho.pdf
3. Voronov, Th.: De Rham cohomology. Lecture notes University of Manchester 2010-2011. Available at
   http://www.maths.manchester.ac.uk/~tv/Teaching/
   Differentiable%20Manifolds/2010-2011/8-cohomology.pdf
4. Weintraub, S.: Differential Forms. Second Edition. Academic Press 2014
5. Andrews, B: Differential forms. Lectures Australian National University. No year. Available at

```
http://maths-people.anu.edu.au/
~andrews/DG/DG_chap13.pdf
```
6. Avramidi, I.: Notes on Differential Forms. Lectures New Mexico Tech 2003. Available at
```
http://infohost.nmt.edu/~iavramid/notes/diffforms.pdf
```
7. Clelland, J.N.: Differential Forms. Lectures University of Colorado at Boulder. No year. Available at
```
https://math.colorado.edu/~jnc/lecture1.pdf
```
8. Bryant, R.L., Chern, S.S., R.B. Gardner, H.L. Goldschmidt, Griffiths, P.A.: Exterior Differential Systems. Springer 2011. Available at
```
http://library.msri.org/books/Book18/
MSRI-v18-Bryant-Chern-et-al.pdf
```
9. Dray, T.: The Hodge Dual Operator. Lecture notes Oregon State University 1999. Available at
```
http://people.oregonstate.edu/~drayt/
MTH437/handouts/dual.pdf
```
10. Gupta, S.N.: Theory of Longitudinal Photons in Quantum Electrodynamics. Prc. Phys. Soc. A 63, 681-691 (1950).

11. Bleuler, K.: A new method for the treatment of the longitudinal and scalar photons. Helv. Phys. Acta 23. 567-586 (1950). Available at
```
http://web.science.mq.edu.au/~dalew/thesis/
Bleuler_HelvPhysActa_23_pp567-586_1950.pdf
```
12. Strocchi, F., Wightman, A.S.: Proof of charge selection rule in local relativistic quantum field theory. J. Math. Phys. 15, 2198-2224 (1974).

13. Bongaarts, PJ.M.: Maxwell's equations in axiomatic quantum field theory.

I. Field tensor and potentials. J. Math. Phys. 18, 1510-1516 (1977).

II. Covariant and non-covariant gauges. J. Math. Phys. 23, 1881-1898 (1982).

14. Bongaarts, P.: The quantum Maxwell field revisited. In preparation. No year.

10 ALGEBRAIC QUANTUM FIELD THEORY

10.1 Introduction

Wightman's theory was the first step on the long road to a rigorous mathematical basis for relativistic quantum field theory, a road with even now no end in sight, although in its more or less mathematically heuristic form, it has been used very successfully in elementary particle physics. As such Wightman's formulation remained close to the main notions of standard field theory, with field operators acting in a Hilbert space of state vectors as the basic elements.

The relation between fields and physical observations is however fairly complicated. For this reason the more direct physical but also more general notion of *local observable* was introduced by Rudolph Haag, and appeared in full bloom in the paper he wrote in 1964 with Daniel Kastler [1].

Wightman's approach was not algebraic but it was later reformulated in an algebraic manner, as we have shown in the preceding chapter. Algebras were from the beginning essential in what became known as AQFT (the abbreviation used by the 'conoscenti' for *algebraic quantum field theory* in the spirit of Haag-Kastler). From our point of view it is, after the algebraic reformulation of Wightman theory, a further example of an *algebraic covariance system* to be treated in this book. So far it is more or less the point that has at present been reached in the long and difficult road to a mathematically rigorous, physically non-trivial quantum field theory. This chapter is devoted to it.

It will be formulated, again, along the lines of Chapter 2, in four axioms, following in the main the material from the books by Haag [2], Araki [3] and Horuzho [4]. See further [5], [6] (both on internet), [7], [8] (on internet) [9]. In the last section of this chapter, a generalized form of AQFT will be presented – or at least suggested – which is needed for incorporating the Maxwell quantum field.

10.2   Observables

The observables form a C^*-algebra $\mathcal{A}$, generated by a system of C^*-algebras $\mathcal{A}(\mathcal{O})$, for every bounded open set $\mathcal{O}$ in 4-dimensional spacetime $\mathbb{R}^4$. The total algebra $\mathcal{A}$ is the norm closure of the union $\cup_{\mathcal{O}} \mathcal{A}(\mathcal{O})$. The selfadjoint elements of $\mathcal{A}(\mathcal{O})$ are called the *local observables*, those of $\mathcal{A}$ the *quasi-local observables* of the theory under consideration.

The system $\cup_{\mathcal{O}} \mathcal{A}(\mathcal{O})$ has to satisfy the following axioms:

1. *Isotony*: $\mathcal{O}_1 \subset \mathcal{O}_2 \;\longrightarrow\; \mathcal{A}(\mathcal{O}_1) \subset \mathcal{A}(\mathcal{O}_2)$.

2. *Lorentz covariance*: There is a representation of the Poincaré group by $*$-autormorphisms $\varphi_{(a,\Lambda)}$ in $\mathcal{A}$, such that

$$\varphi_{(a,\Lambda)}(\mathcal{A}(\mathcal{O})) = \mathcal{A}((a,\Lambda)\mathcal{O}),$$

with

$$(a,\Lambda)\mathcal{O} = \{x_1 \in \mathbb{R}^4 \mid x_1 = \Lambda x + a, \; \forall x \in \mathcal{O}.$$

3. *Locality*: For every space-like related $\mathcal{O}_1$ and $\mathcal{O}_2$, i.e. with

$$x^\mu y_\mu = x^0 y^0 - x^1 y^1 - x^2 y^2 - x^3 y^3 < 0, \; \forall x \in \mathcal{O}_1, y \in \mathcal{O}_2,$$

which means that all elements of $\mathcal{A}(\mathcal{O}_1)$ and $\mathcal{A}(\mathcal{O}_2)$ commute pairwise. This is, of course, a typical relativistic property. It forbids time travelling (tachions).

10.3   States

These are, as was stated in Chapter 2, Section 2.2, positive normalized linear functionals ω on $\mathcal{A}$. This means:

1. $\omega(a^*a) \geq 0$, for every a in $\mathcal{A}$,
2. $\omega(1_\mathcal{A}) = 1$.

Problem. Show that the positivity of ω implies its continuity.

A physically admissible state ω should be *Poincaré invariant*, i.e.

$$\omega(\varphi_{(\Lambda,a)}b) = \omega(b),$$

for all Poincaré transformations (Λ, a), for all b in $\mathcal{A}(\mathcal{O})$ and all $\mathcal{O}$. One can show that this implies that the representation of the Poincaré group by $*$-automorphisms $\varphi_{(\Lambda,a)}$ of the algebra $\cup_{\mathcal{O}}\mathcal{A}(\mathcal{O})$ is *unitarily implementable* in the GNS representation Hilbert space $\mathcal{H}_\omega$, meaning that there are unitary operators $U_\omega(a,\Lambda)$ in $\mathcal{H}_\omega$ with the properties

1.

$$U_\omega(a,\Lambda)(\Omega_0)_\omega = (\Omega_0)_\omega, \; \forall(a,\Lambda),$$

and with $(\Omega_0)_\omega$ the ground state (0-particle state, vacuum state) of the system.

2. For all elements (a, Λ) of the Poincaré group, the unitary operators $U(a, \Lambda)$ act in $\mathcal{H}_\omega$, implementing the automorphisms $\varphi_{(a,\Lambda)}$, as

$$U(a, \Lambda)\pi_\omega(\varphi_{(a,\Lambda)}(b))U(a, \Lambda)^{-1} = \pi_{\omega_{(a,\Lambda)}}(\varphi(b)), \quad \forall b \in \mathcal{A}.$$

10.4 Interpretation

Following Chapter 2, Section 2.3, one constructs for a given state ω, the Hilbert space $\mathcal{H}_\omega$ and the representation π_ω from $\mathcal{A}$ into the space $\mathcal{B}(\mathcal{H}_\omega)$, the space of bounded operators in $\mathcal{H}_\omega$. In this an element b from $\mathcal{A}$ is represented by an operator $\pi_\omega(b)$ in $\mathcal{H}_\omega$. There is a uniquely defined unit vector ψ_ω. This is the *GNS construction*, fully described in Chapter 7.

The expression $\omega(a) = (\psi_\omega, \pi_\omega(a)\psi_\omega)$ is the *expectation value* $\overline{a}_\omega$ of the observable a in the state ω in the usual sense of quantum theory. From this higher moments can be derived, for instance the second moment leading to the standard deviation of the observable a in the state ω. One may also determine the full probability distribution.

It should be noted that in standard probability theory any stochastic variable has a distribution function, but higher moments may not always exist. This remains true for this 'quantum probability' situation.

10.5 Symmetries

Algebraic quantum field theory is a relativistic theory, so the meaning of time development is not an intrinsic one but depends on the coordinate system that is being used. That leaves us with symmetry, in fact symmetry with respect to the Poincaré group. See for this the mathematical intermezzo - *Poincaré or inhomogeneous Lorentz group*, to be found in Chapter 2, Section 6. So AQFT is an algebraic dynamical system.

The theory predicts a separate discrete symmetry, CPT, just as in Wightman the ory, discussed in Chapter 8, and experimentally confirmed. This is the only non-trivial statement with this property. Anyway, both theories have great beauty and give an attractive and convincing general theoretical background for particle physics. But apart from this, AQFT is – for its practitioners at least – 'Art Pour l'Art'.

For the formulation of explicit examples, the GNS representation is important, in particular for *super-selection rules* which are a common feature of AQFT.

- *Physics intermezzo*.

Super-selection rules. Selection rules in quantum theory restrict transitions between states. Often such transitions are only forbidden in lowest order. There are also absolute selection rules. The principal example is a *super-selection rule*. The Hilbert space of states is then the direct sum of so-called super-selection sectors. A superposition of state vectors from different sectors has no physical meaning. There is such a super-selection rule between particles with integer spin (bosons) and particles with half integer spin (fermions). Since the mid-seventies supersymmetry, which would connect bosons and fermions, has been discussed as a possibility for breaking this super-selection rule. But this phenomenon has up to now not been detected experimentally. For a good introduction to super-selection rules, see [10] (on internet).

- End of physics intermezzo

10.6   A general problem

Typical standard books on AQFT like [2] do not mention the Maxwell field at all, clearly because it does not fit in the framework of AQFT as it has been developed over the years, even though the Maxwell field was not only the first field that was quantized, but, as a part of quantum electrodynamics, was and still is the most successful quantum field theory we have.

In the preceding chapter we noted that the same holds for Wightman theory. There the problem is clear. The requirement of manifest Poincaré invariance of the vector potential $A_\mu(x)$ leads inevitably to 'pseudo-Hilbert spaces', i.e. spaces with an indefinite inner product. Also algebras more general than C^*- and von Neumann algebras, extensively studied in mathematics, but rarely or not at all used in mathematical physics, have a role to play. Because of this AQFT is clearly in need of a generalization. In this section we shall propose a sketch for this.

Our point of departure is the free Maxwell field, as described in the preceding chapter within the Wightman approach, a linear theory which can be analyzed completely. We may therefore use the results obtained there as an example of the general set up for all algebraic quantum field theories, that can be expected to exhibit the same problem, such as, for instance all theories that involve gauge transformations, theories of the Yang-Mills type. Most of the physically interesting theories that describe systems of interacting particles in high-energy physics (elementary particle physics) are of this type.

We begin with the unphysical fields, which for the Maxwell field are the local 'observables' generated by the vector field $A_\mu(x)$, and in the general case a system of fields, to be denoted as $A_{...}(x)$. Because they do not have a direct physical meaning, one may speak of 'pseudo-observables'. They cannot be omitted, however, because they are essential for the formulation of the dynamics of the theory, in particular when the Lagrange formalism is used. They take care, so to speak, of the inner mechanism of the theory. For the local pseudo-observables, we postulate a $*$-algebra, to be denoted as $\mathcal{A}^{\text{non-ph}}$. In the case of the algebraic Wightman approach this algebra was a tensor algebra over a Schwartz testfunction space, which have been studied extensively. For the more general situation of AQFT we need a broader class of locally convex algebras. See for this the mathematical appendix - *Locally convex spaces and algebras*, which can be found in Section 5 of Chapter 6. This is the minimal requirement; the world of more general topological vector spaces is a desert. Determining and studying the appropriate type of locally convex algebra for generalized AQFT is the great challenge, which is waiting to be met in the future.

A state of the unphysical system is a generalization of the notion of state in AQFT, defined earlier in this chapter. Positivity is dropped; instead we only require continuity, which would be implied by positivity. This state has to be Poincaré invariant. It determines a GNS representation space $\mathcal{H}_\omega^{\text{non-phys}}$, a pseudo-Hilbert space, because it has a continuous non-degenerate inner product $(\cdot, \cdot)_\omega^{\text{non-phys}}$, The physical observables can be derived from the unphysical ones. For the Maxwell field situation we have equation 10.1.3 in Chapter 10. For the general case there will be a similar connection. The result is that we have a pseudo-Hilbert space in which both the unphysical and the physical pseudo-observables act. Because of the Poincaré invariance of the state the group of Poincaré automorphisms in $\mathcal{A}^{\text{non-ph}}$, these automorphisms are inplemented by pseudo-unitary operators in $\mathcal{H}_\omega^{\text{non-phys}}$.

This completes the 'unphysical' part of a generalized AQFT.

Next the 'physical' part of the formalism. In the unphysical GNS representation space there is a null space $\mathcal{N}_\omega^{\text{non-phys}}$, due to the fact that the inner product in $\mathcal{H}_\omega^{\text{non-phys}}$ is not positive definite. The true physical Hilbert space is the quotient space $\mathcal{H}_\omega^{\text{phys}} = \mathcal{H}_\omega^{\text{non-phys}}/\mathcal{N}_\omega^{\text{non-phys}}$. It can be completed to a Hilbert space. (See for the notion of quotient space the mathematical intermezzo - *Quotient spaces*, to be found in Section 2 of Chapter 7). The physical algebra $\mathcal{A}^{\text{phys}}$ is the quotient algebra over a two-sided ideal in $\mathcal{A}^{\text{non-phys}}$. (For a definition of a two-side algebra ideal, see the mathematical intermezzo - *Quotient spaces*, to be found in Section 2 of Chapter 7). The pseudo-unitary operators that represent the Poincaré group in $\mathcal{H}_\omega^{\text{non-phys}}$ descend to unitary operators in $\mathcal{H}_\omega^{\text{phys}}$, which means that only in the physical picture there is both manifest Poincaré covariance and a positive definite inner product, which is, of course, sufficient.

This completes our sketch of a generalized AQFT, needed for all Yang-Mills type quantum field theories. Much further research has to be done, in particular on an appropriate choice of the locally convex algebras to be used, and on their precise mathematical properties.

10.7 Final remarks

The description of a physical system with as its basic notion the algebra of its observables, emerged in a quantum mechanical environment, in the

96

classic paper by Irving Segal in 1947 (Reference [1] of the Preface). Its
ideas were in 1964 applied to quantum field theory in a paper by Rudolph
Haag and Daniel Kastler (Reference [2] of the Preface). Much later, in
2009, an explicit presentation of both classical and quantum systems in this
framework was sketched in a short and elegant book of Ludwig Faddeev and
O.A. Yakubovskiĭ (Reference [3] of the Preface). In the present book this
idea is developed further, with an emphasis on presenting pairs of similar
theories, one classical, the other from quantum theory. This is sometimes
far-fetched, for instance in the case of classical mechanics, as was observed
in Section 2 of Chapter 3.

There are no obvious classical versions of relativistic quantum field theories, either in the Wightman sense or in the sense of AQFT. However,
there are non-relativistic systems which can – in a vague way – play this
role, for instance, the infinite classical Ising spin systems, in which one can
look for phase transitions. It is known that the 1-dimensional case does
not have phase transitions; the 2-dimensional model does have one, as was
proved by Lars Onsager in 1944. See [11] (on internet).

References

1. Haag, R., Kastler, D.: An Algebraic Approach to Quantum Field Theory. J. Math. Phys. **5**, 848-861 (1964).
2. Haag, R.: Local Quantum Physics (2nd ed.). Springer 1992.
3. Araki, H.: Mathematical Theory of Quantum Fields. Oxford 2009.
4. Horuzhy, S.S.: Introduction to Algebraic Quantum Field Theory. Springer
1990.
5. Fredenhagen, K., Rehren, K-H.: Algebraic Quantum Field Theory. University of Göttingen Lecture Notes 1998. Available at
ftp://physik.uni-goettingen.de/pub/
papers/rehren/99/AQFT-en.pdf
6. Buchholz, D.: Current Trends in Axiomatic Quantum Field Theory.
Talk at 'Ringberg Symposium on Quantum Field Theory 1998. Available
at
https://cds.cern.ch/record/372563/files/9811233.pdf
7. Dybalski, W.: Algebraic Quantum Field Theory. München Technical
University Lecture Notes. No year. 8. Halvorson, H.: Algebraic Quantum
Field Theory. In "Handbook of the Philosophy of Physics". Butterfield, J.,
Earman, J., (eds). North Holland 2006. Available at
https://www.princeton.edu/~hhalvors/aqft.pdf
9. Roberts, J.: More Lectures on Algebraic Quantum Field Theory. Göttingen
University. No year.
10. Giulini, D.: Superselection Rules. Max Planck Institute Potsdam. No
year. Available at
https://core.ac.uk/download/pdf/11921700.pdf
11. Cipra, B.A.: An Introduction to the Ising Model. Lecture notes. St.

Olaf College, Minnesota. No year. Available at
    http://www.ww.amc12.org/sites/default/files/pdf/
    upload_library/22/Hasse/00029890.di991727.99p0087h.pdf

11 NON-COMMUTATIVE STOCHASTIC GRAPH THEORY

11.1 Introduction

In probability theory there is a recent subject of great interest. It is usually called *stochastic graph theory* or *random graph theory*. In this chapter we discuss this together with the non-commutative version, which we shall call *non-commutative stochastic graph theory* or *quantum stochastic graph theory*, and on which there is so far very little or nothing to be found in the existing literature, although it is a fairly obvious generalization of the 'classical' situation. In the first case we have a special type of classical probability theory, with a system of graphs as its sample space Ω. In the quantum case there is only a heuristic notion of a sample space, which is helpful to get suggestions for the rigorous theory. It is a 'non-commutative sample space', in the spirit of our broad discussion of non-commutative geometry in Chapter 4. In a rigorous sense we have a non-commutative algebra, generated by the elements of the graph.

- Mathematical intermezzo.

Graph theory. A *graph* consists of a non-empty (at most countably infinite) set of points p_j (*vertices* or *nodes*) together with a non-empty set of lines $(i,j), i \neq j$ (*edges*) between these points. Two points can be connected by only a single line, so $(i,j) = (j,i)$. A graph is connected if every pair of points can be connected by a *path*, i.e. a series of lines. In a *directed* graph the lines have a direction; (j,i) goes from p_i to p_j. Two points can only be connected by a single path with the same direction. In this case $(i,j) \neq (j,i)$. The simplest sort of a (non-directed) graph is a *tree*. It consists of a single connected component and does not contain *cycles*, lines by which one can return to an initial point, like the sequence $(i,j), (j,k), (k,l), (l,i)$.

Graph theory is an example of discrete mathematics and is a part of combinatorics. Its basis was laid by Leonhard Euler in the 18th century. It has a wide range of applications, in mathematics, e.g. combinatorics, in particle physics, in particular the calculation of the integrals connected with Feynman diagrams, in applied physics with for example the study of electric circuits, biology with the study of the neuron structure of the brain, in the social sciences, in signal processing, artificial intelligence. In citation analysis, in which one studies the propagation of scientific knowledge, by looking at how and how frequently authors cite work of predecessors. Note that in this case one has to use directed tree graphs. Graph theory provides a unified framework for an approach to a variety of dynamical systems.

In this chapter we focus on graph theory as a basis for probabilistic problems, in particular for the study of stochastic processes, first for the classical situation, and then for its 'non-commutative' or 'quantum' generalization.

Good books on graph theory are [1], [2], [3] (the last complete on internet). On applications of graph theory [4] (on internet), [5].

- End of mathematical intermezzo

- Mathematical intermezzo.

Stochastic processes. We described in the mathematical intermezzo *probability theory* - to be found in Chapter 3, Section 3, a probability space as a triple $(\Omega, \mathcal{F}, P)$, with Ω the sample space, $\mathcal{F}$ a system of subsets of Ω and P a probability. A stochastic variable is then a measurable function on Ω, real-valued or integer-valued; with P defining a probability distribution on this variable. A stochastic process is a system of time-dependent random variables $X_t, t \in \mathbb{R}$ (continuous time), or $X_n, n \in \mathbb{N}$ (discrete time). There are joint probability distributions for all finite sets of X_n. A process is *independent* iff the joint probability distributions are products of the distribution functions of the single stochastic variables.

A discrete stochastic process is sometimes called a *time series*. Markov processes form an important class of stochastic processes. They are characterized by the property that the probability distribution at a time t_b depend on the distribution at an earlier time t_a but *not* on a distribution at any still earlier time. They have, so to speak, no memory. Discrete Markov processes are usually called *Markov chains*. A simple but important example of a Markov chain is the 1-*dimensional random walk*. This starts at $t = 0$ with a point distribution concentrated on the origin, and then proceeds in single steps, with equal probability, to the left or to the right. *Brownian motion*, the random motion of small particles suspended in a liquid, first seen through a microscope by the botanist Robert Brown in 1827, is an example of a continuous 3-dimensional random walk. It was explained and given a precise mathematical description by Albert Einstein in 1905.

Good introductions to the theory of stochastic processes are [6] and [7] (both on internet). On Markov processes [8] (on internet).

- End of mathematical intermezzo

11.2   Stochastic graph theory

In what is called in the literature 'stochastic graph theory' or 'random graph theory' one studies probabilistic models with as sample space a finite or countably infinite system of graphs. A good example is the work of Den Hollander et al. [9] (on internet). It is a generalization of classical statistical mechanics. As sample space the authors choose instead of the classical phase space of a system of particles a collection of graphs. By imposing additional conditions they obtain the analogues of the ensembles of classical statistical mechanics, in particular the micro-canonical and the canonical ensemble. This approach has various applications, inside and outside physics, in biology, for example. See [10] (on internet).

11.3   Quantum stochastic graph theory

In the literature 'quantum stochastic graph theory', or just 'quantum graph' theory, usually stands for a theory in which graphs are provided with a differential operator, for example the Hamiltonian operator of a system of particles. See [11] (complete monograph on internet).

In this section we present a different idea which we shall also denote as 'quantum stochastic graph theory', or alternatively as 'non-commutative stochastic graph theory'. So far, to the best of our knowledge, this has, as yet, not been discussed in the literature, but it seems however quite natural and it fits very well in the context of this book, in which the main idea is that classical and quantum physics can be treated on the same footing, by taking as basic notion the algebra of observables, commutative for the classical case, and non-commutative for the quantum situation.

The pertinent example is the well-known pair classical and quantum probability theory. The standard formulation of the first, due to Kolmogorov, is by means of a probability space $(\Omega, \mathcal{F}, P)$, with Ω the sample space of events, $\mathcal{F}$ a system of sets in Ω, and P the probability of the events. This leads to the distribution functions of all observables. See the mathematical intermezzo *probability theory* - to be found in Chapter 3, Section 3.

There is an equivalent algebraic reformulation for this. The almost everywhere bounded measurable functions on Ω form a commutative von Neumann algebra $\mathcal{A}$ which can be uniquely extended to the system of observables $\mathcal{F}$, the probability P defines a normed positive functional ω on $\mathcal{F}$. The original classical formulation can be recovered from the pair $(\mathcal{A}, \omega)$ by means of the GNS construction, discussed in Chapter 6. For the equivalence of probability spaces and commutative von Neumann algebras, see [12] (Prop. 1.81.1) (complete monograph on internet). See also for the mathematical intermezzo *von Neumann algebras* - to be found in Chapter 5, Section 4.

The quantum (or non-commutative) analogon is obtained by deforming $\mathcal{A}$ to a non-commutative von Neumann algebra $\mathcal{A}^q$.

In general there are many such deformations possible, all leading to different physical models. In the example of classical mechanics the algebra $\mathcal{A}$ is generated by functions of the canonical variables p and q. From this a well-known procedure leads to one of the possible canonical quantizations, in which the p and q become non-commuting algebraic objects.

Quantum probability theory has become an important field on which a vast literature exists. See for an excellent introduction [Maassen] (on internet). By probability theory we meant so far in this book *classical probability theory*. See the mathematical intermezzo *Probability theory* - appearing in Chapter 3, Section 3. In the light of the discussion in this chapter it would be more appropriate to reserve the term probability theory for a general notion, containing classical and quantum probability as subnotions.

There is a rather surprising omission in the existing literature on quantum probability, papers and lecture notes. All authors on this subject seem to be unaware of the simple and transparant algebraic formulation, explained in this book. Instead one usually starts with discussing the obvious probabilistic nature of quantum theory and than introduces the idea of quantum probability by describing a number of explicit physical models, of increasing complexity, everything usually in a very competent manner, but completely obscuring the essential simplicity of the underlying mathematical structure and the way quantum probability fits in the general scheme of quantum theory. See for instance [13] and [14] (both on internet). In Maassen's very clear lecture notes [14] this algebraic idea is at least discussed in the beginning, but not further developed in the remainder of the notes. This is probably because the explicit models he discusses are spin models, fermionic quantum systems, which do not quite fit in the standard algebraic dynamical systems framework. This was discussed in Chapter 2, Section 7.

The algebraic approach leading to a non-commutative situation can be illustrated by starting from, for example, Den Hollander's model. This situation is discrete. For every graph in Ω there is a so-called simple function, 1 on this graph and 0 elsewhere. These functions generate a commutative algebra – it is enough to take for this all linear combinations of these simple functions. This algebra can be extended to the commutative von Neumann algebra $\mathcal{A}$ of bounded observables. The probability P defines again a normed positive functional on $\mathcal{A}$.

The next step is defining suitable non-commutative deformations $\mathcal{A}^q$ of $\mathcal{A}$. For this the graphs, and therefore the simple functions characterizing the graphs, have to be characterized in a unique manner. This can be done by *incident matrices*. See [15] (on internet). It leads to a system of 'abstract' elements, that can be used to define a deformed version $\mathcal{A}^q$ of $\mathcal{A}$. There are again different possibilities for such a deformation.

In the case in which the number of graphs is N, there are $\mathbb{R}^N$ formal products of the symbols for the N graphs. The algebra $\mathcal{A}^{\text{univ.}}$, generated by these formal objects is the largest algebra containing those elements, and as such an universal object. Is is of course not an algebra which can be used in applications. We need to construct from it quotient algebras, by choosing suitable algebra ideals. In such an algebra there will be many commuting pairs of elements. An obvious example is Den Hollander's commutative algebra, as such not explicitly mentioned by him, as he does not use our algebraic framework.

Consider quantum statistical mechanics, seen as an example of quantum probability, as discussed above. All the observables generated by the p's commute with all the observables generated by the q's. The 'physical' algebra $\mathcal{A}^q$ is much smaller than the universal algebra $\mathcal{A}^{\text{univ.}}$.

The particular version of a quantum stochastic graph theory, put forward here, has so far not been discussed in the literature. Its detailed properties remain to be investigated. This is even more true for its possible applications, although a subject like quantum information comes to mind in this respect. The wide range of applications of stochastic graph theory may be expected to have 'quantum' versions.

11.4   Examples

11.4.1   The quantum Black-Scholes equation

The Black-Scholes equation is a stochastic equation in financial mathematics used for predicting the price of options.

Irving Segal, the originator of the algebraic view of quantum mechanics, and William Segal, an econometrist, probably his son, proposed a "quantum Black-Scholes equation" [17] (on internet).

The basis of the idea of Segal & Segal is the following. Consider the classical description of a one-dimensional particle. All observables are at the same time exactly measurable. A quantum description of the same situation is more precise. Position and momentum are no longer at the same time exactly measurable. One has Heisenberg's inequality prohibiting this. Segal & Segal observe that the pricing of options shows extreme irregularities that cannot be described by the standard Black-Scholes theory. They suggest that their quantum Black-Scholes theory is better suited to this.

The notion of a quantum Black-Scholes equation has been widely discussed. See [18] (on internet).

11.4.2   Other examples

Other realizations of the same idea are quantum random walks and quantum Brownian motion. We shall not discuss this here, but see [19] and [20] (both on internet).

References

1. Diestel, R.: Graph Theory. Springer 2012.
2. Wilson, R.J.: Introduction to Graph Theory. Pearson 2012.
    3. Bondy, J.A., Murty, U.S.R.: Graph Theory. Springer 2008. The complete book can be found at
    `https://www.classes.cs.uchicago.edu/archive/`
    `2016/spring/27500-1/hw3.pdf`
4. Estrada, F.: Graph and Network Theory. University of Strathclyde, Glasgow lecture notes. Accessible at
    `https://arxiv.org/pdf/1302.4378.pdf`
5. Schmidt, D. et al.: Random Graph and Stochastic Process Contributions to Network Dynamics. Ohio State University and University of Iowa lecture notes. No year.
6. Breuer, L.: Introduction to Stochastic Processes. University of Kent Lecture notes. No year. Accessible at
    `https.www.kent.ac.uk/smsas/personal/lb209/files/sp07.pdf`
7. Corcuera, J.M.: A Course in Stochastic Processes. University of Barcelona lecture notes. No year. Accessible at
    `http://www.ub.edu/plie/`
    `personal_PLiE/corcuera_HTML/Teoria4.pdf`
8. NN : Markov Processes with Countable State Spaces. M.I.T. lecture notes. 2011. Available at
    `http://www.rle.mit.edu/rgallager/documents/6.262vcpw7.pdf`
9. Den Hollander, F. : Breaking of ensemble equivalence in complex networks. Leiden LCN2 network 2015. To be obtained at
    `http://www.isibang.ac.in/~athreya/`
    `pcm/pdf.php?file=CMa.pdf`
10. DeVille et al. Dynamics of Stochastic Neuronal Networks and the connections to Random Graph Theory. Math. Model. 2010. Accessible at
    `http://www.math.uiuc.edu/~rdeville/research/`
    `MMNP-DPS-Jan10.pdf`
11. Berkolaiko, G.: An Elementary Introduction to Quantum Graphs. Arxiv preprint [math-ph] 2016. Accessible at
    `https://arxiv.org/pdf/1603.07356.pdf`
12. Sakai, S.: C^*-Algebrs and W^*-Algebras. Springer 1971. The complete book is available at
    `http://alirejali.ir/afiles/up/other/book1/`
    `C-%20algebras%20and%20W-algebras1.pdf`

13. Accardi, L. : Topics in Quantum Probability. Lecture notes Istituto di Cibernetica, Napels 1981. Available at
   `https://art.torvergata.it/retrieve/handle/`
   `2108/82347/162180/Ac81a_Topics`
   `%20in%20quantum%20probability.pdf`
14. Davies, E.B., Lewis, J.T. : An Operational Approach to Quantum Probability. Commun. math. Phys. 17, 239-260 (1970). Available at
   `http://projecteuclid.org/download/pdf_1/`
   `euclid.cmp/1103842336`
15. Maassen, H. : Quantum Probability Theory. University of Nijmegen lectures 1998. Available at
   `https://www.math.ru.nl/~maassen/lectures/qp.pdf`
16. NN : Graph Matrices. Lecture notes. No year. Available at
   `http://compalg.inf.elte.hu/~tony/Oktatas/`
   `TDK/FINAL/Chap%2010.PDF`
17. William Segal, I.E. Segal: The Black-Scholes pricing formula in the quantum context. Proc. Natl. Acad. Sci. USA, 95, 4072-4075 (1998). Available at
   +http://www.pnas.org/content/pnas/95/7/4072.full.pdf+
18. Luigi Accardi, Andreas Boukas: The Quantum Black-Scholes Equation. Preprint of the Centro Volterra, Rome, and Department of Mathematics and Natural Sciences, Athens. 2007. Available at
   +https://arxiv.org/pdf/0706.1300.pdf+
19. S. Attal et al. Open Quantum Random Walks. Preprint Lyon University. No date. Available at
   +http://math.univ-lyon1.fr/ attal/Mesarticles/OQRW2.pdf+
20. László Erdos: Lecture Notes on Quantum Brownian Motion. preprint University of Munich. 2010. Available at
   +http://www.mathematik.uni-muenchen.de/ lerdos/+
   +Notes/houchnotes.pdf+

Mathematical Intermezzos

1. Alphabetic list

The number behind each item refers to the chapter and section where it can be found. Example: 3.2 means Chapter 3, Section 2.

2. List in order of appearance

Relations between various topics

There are hierarchical relations between the mathematical intermezzos. Some of these relations are shown in the following schemes.

a.

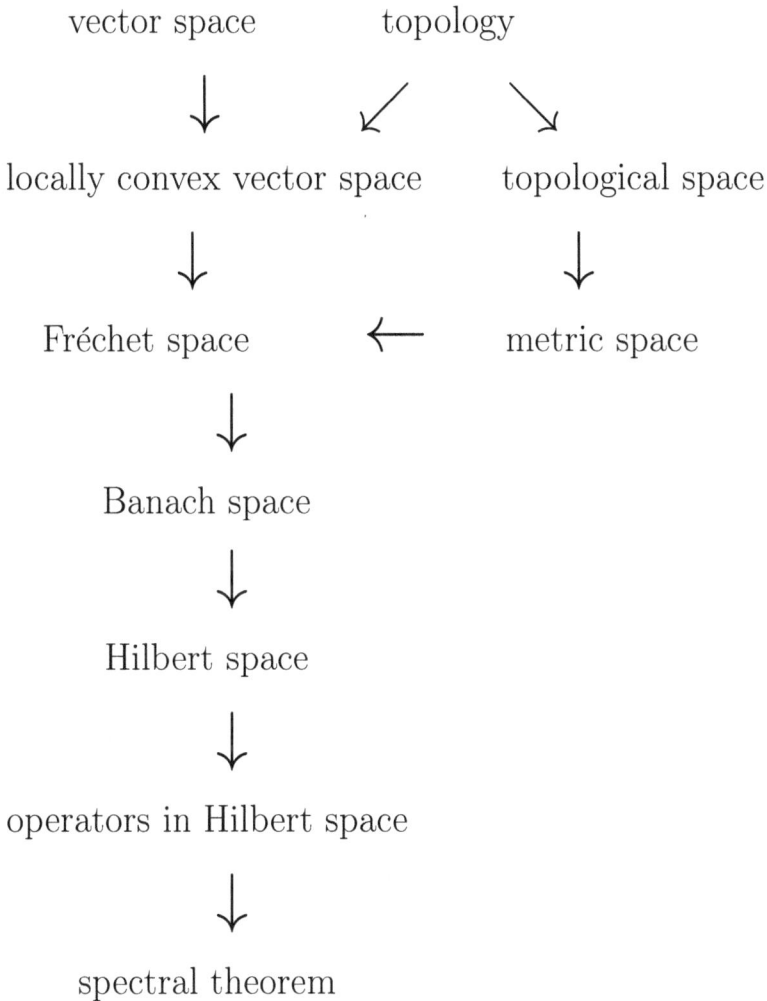

b.

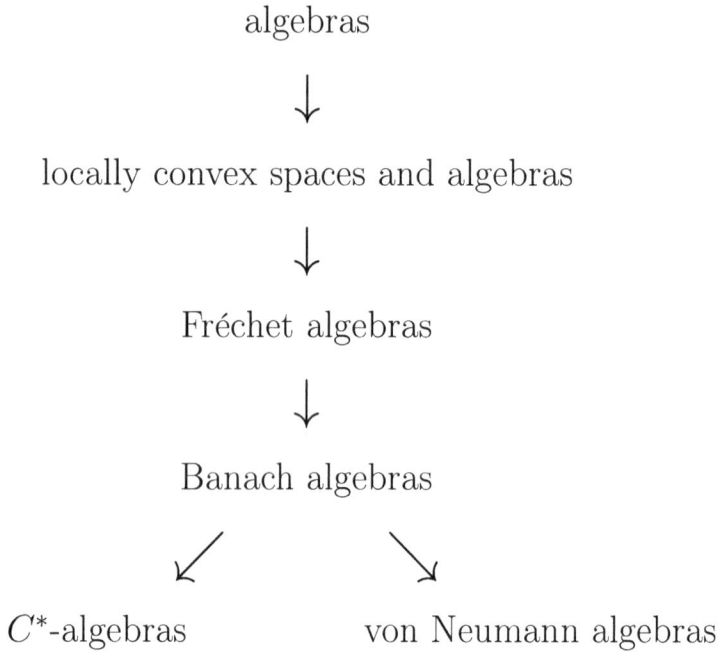

algebras

↓

locally convex spaces and algebras

↓

Fréchet algebras

↓

Banach algebras

↙   ↘

C^*-algebras   von Neumann algebras

c.

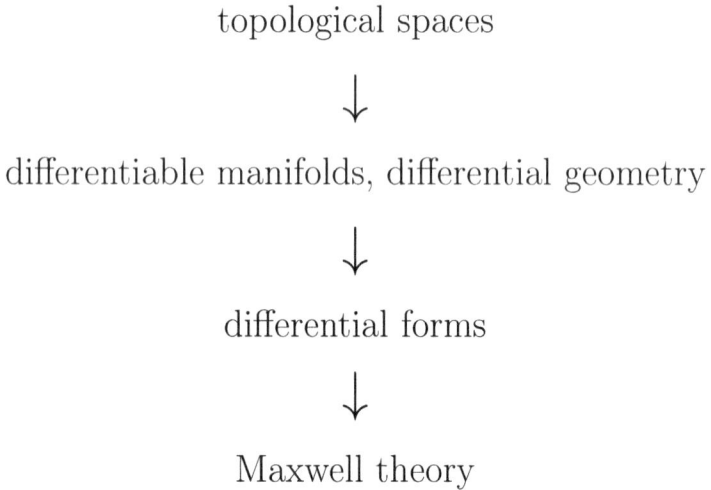

topological spaces

↓

differentiable manifolds, differential geometry

↓

differential forms

↓

Maxwell theory

d.

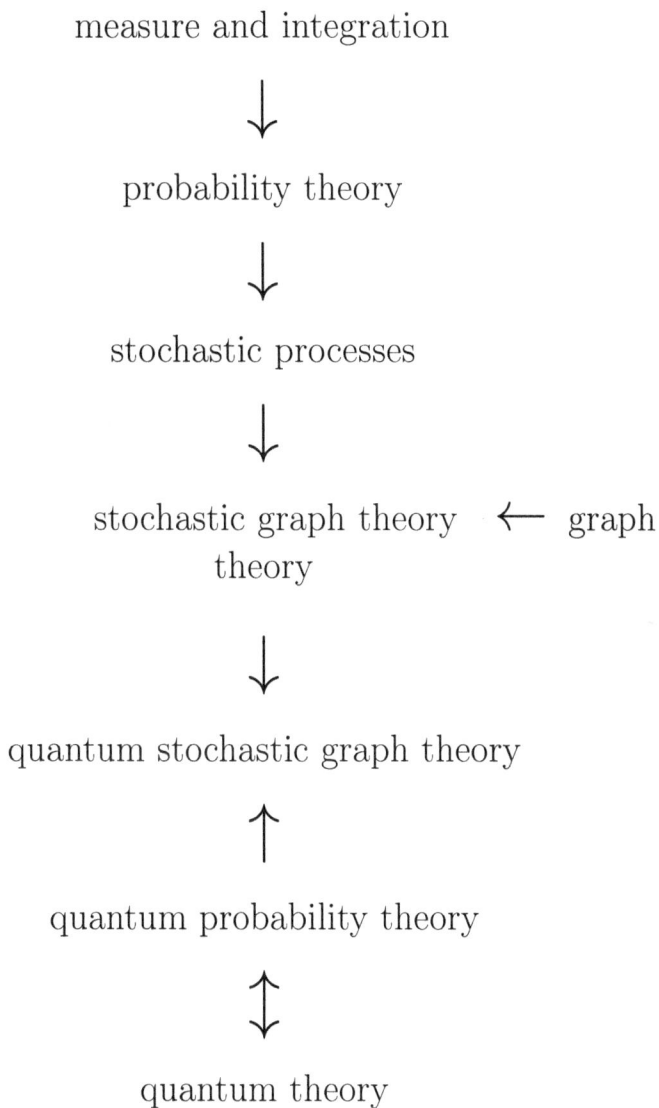

measure and integration

↓

probability theory

↓

stochastic processes

↓

stochastic graph theory ⟵ graph theory

↓

quantum stochastic graph theory

↑

quantum probability theory

↕

quantum theory

List of authors cited

After each name, or names, there is a reference to the full annotated bibliographies appearing at the end of most of the chapters. Only the surnames of the author(s), without first names or initials, together with the year of publication, appear in this list. After this a number, like for example, 12, referring to the Reference section of Chapter 12, where complete bibliographic information can be found on the item in question, with additional information, e.g. initials or full first names, just as given by the authors, and above all on internet availability. In this P refers, of course, to the reference section of the Preface. In case of more than two authors, we write the name of the first author, followed by "et al.".

A

Accardi. 1981 - 11, 2007 - 11
Alabiso, Weiss. 2015 - 3
Andrews. No year - 9
Anevski. 2012 - 3
Araki, 2009 - 10
Armstrong, 1983 - 4
Arveson. 1996 - 3, 2002 - 3, 1976 - 5
Assche, van, et al.. No year - P
Attal et al. No date - 11
Avramidi. 2011 - 2, 2000 - 3, 2003 - 9
Azmi. 2012 - 4

B

Baggot. 2014 - P
Ballentine. 1998 -3
Besnard. 2013 - 4
Berberian. 2013 - 8
Berkolaiko. 2016 - 11
Bertlmann. No year - 3
Binney. 2013 - 3
Blackader. 2013 - 5
Bleuler. 1950 - 9
Bogolubov et al..1990 - 7
Bondy, Murty. 2008 - 11
Bongaarts. 2014 - P, 2004 - 4, 1977, 1982 - 9, No year - 9
Bongaarts et al.. 1994 - 2
Borchers. 1962 - 8
Boukas - 2007 - 11
Brattelli. 2003 - 3

Brattelli, Robinson. 2003 - 3
Breuer. No year - 11
Bryant et al.. 2011 - 9
Buchholz. 1998 - 10
Buchholz et al.. 2015 - 6

C

Cherney, et al.. 2013 - 2
Chung. 2000 - 3
Cipra. No year - 10
Clelland. No year - 9
Connes. No year - 4, 1994 - 4
Corcuera. No year - 11
Cresser. 2005 - 2

D

Davies, Lewis. 1970 - 11
Dawkins. 2005 - 2
Den Hollander. 2015 - 11
DeVille et al. 2010 - 11
Diestel. 2012 - 11
Dieudonné. 1953 - 5
Dirac. 1927 - 7
Dixmier. 1969 - 5, 1981 - 5
Dobrushin. 1994 - 3
Dray. No year - 7, 1999 - 9
Dudley. 2002 - 3
Durrett. 2010 - 3
Dybalski. No year - 10
Dyson. No year - 3

E

2013 - 6 , No year - 7 (3x), No year - 11

Novak. 2011 - 2

O

Orfanides. No year - 3

P

Penrose. 2004 - P

Peskin. ?????? Powers. 2016 - 3

Pulé. 1984 - 6

R

Renault. No year - 5

Roberts. No year - 10

Rothman. 1999 - 2

Ryder. 1985 - 7

S

Sakai. 1971 - 5, 11

Samuelson. 1990 - 2

Schmidt et al.. No year - 11

Schulz. 2011 - 2

Schwartz. 1967 - 5

Scrinzi. 2012 - 3

Segal. 1947 - P, 1995 - 4

Segal. 1998 - 11.4 2x

Seifert, 1980 - 4

Sitarz. 2013 - 4

Smith. No year - 2

Soper. 2012 - 2, - 3

Sparks. 2015 - 3

Srednicki. 2007 - 7

Streater, Wightman. 2000 - 7

Strocchi, Wightman. 1974 - 9

Strocchi. 2005 - P

T

't Hooft. 2004 - 7

Tao. 2011 - 3

Taylor. 2005 - 3, No year - 3

Teschl. 2000 - 3

Thomas. No year - 4

Tillmann. No year - 9

Timoney. 2008-2009 - 3

Tong. 2006-3007 - 7

U

Uhlmann. 1962 - 8

V

Van den Brand. No year - 7

Van der Waerden. 2007 - 3

Van Enk. 2009 - 3

Vilfan. No year - 3

Vogt. 2000 - 5

von Neumann. 1955 - 2, 3, 7

Van Vliet. 2008 - 3

Voronov. 2010-2011 - 9

Vvedensky. No year - 2

W

Wallach. No year - 2

Walker. 1998 - 2

Warner. 1983 - 4, No year - 5

Weinberg. 2005 - 7, 1997 - 7

Weintraub. 2014 - 9

Weyl. 2003 - 2

Wightman. 1956 - 7

Wilde. No year - 7

Williams. 2014 - 4

Wilson. 2012 - 11

Woit. 2004 - P

Wyss. 1969 - 8

Y

Yeh. No year - 2

Z

Zee. 2003 - 7

Zitkovic. 2013 - 3

Subject index

The number behind each item refers to the chapter and section where it can be found. Example: 3.2 means Chapter 3, Section 2.

The items 'quantum theory', 'quantum mechanics' and 'algebraic dynamical system(s)' are not mentioned in this index.

H

Hamiltonian function   3.4
Hamiltonian operator   2.5, 3.4, 7.2, 11.3
Hausdorff   4.2
Hausdorff topological space   4.2
heat bath   3.2
Heisenberg's matrix mechanics 3.4
Heisenberg's uncertainty principle 3.4
hermitian adjoint   3.5
hermitian operator   3.5
high energy physics   8.4, 10.6
Hilbert space   2.2 (2x), 2.4, 2.6, 3.1, 3.3, 3.4, 3.5 (3x), 3.7, 5.3, 5.4 (3x), 7.3 (2x), 7.4, 7.5, 8.4, 9.4, 10.3, 10.4, 10.6
Hilbert space of physical states   5.4
Hilbert space of state vectors   10.1
Hilbert space of states   10.5
Hilbert state space   3.6, 7.2
higher moments   3.4, 10.4 (2x)
hocus pocus   7.2
Hodge duality   9.3
homogeneous Lorentz group $O(1,3)$ 2.6
hydrodynamics   7.1
hydrogen atom   3.4

I

ideals   11.3
idempotency   9.2
imaginary unit   5.2
implemented   10.6
implementing the automorphisms 10.3
important theorem by Gelfand and Naimark   5.3
improper integrals   3.3
incident matrices   11.3
inclusion   4.2
incoming and outgoing particles 8.4

incorporating the Maxwell field 10.1
independent   11.1
indicator function   3.3
indefinite inner product   P (2x), 6.2, 9.4, 10.6
indefinite inner product space   5.1, 9.1
infimum   3.3
infinite classical Ising spin system 10.7
infinite counter terms   7.2
infinite dimensional   3.5
infinite dimensional case   3.5
infinite-dimensional integrals   2.7
infinite-dimensional vector space 2.2
infinite reservoir   3.2
infinite sequence   4.2
infinite system of seminorms   5.3, 5.4
infinitely differentiable functions 4.4, 8.2
infinities   7.2, 7.4
inhomogeneous Lorentz group   2.5, 2.6, 3.6, 7.4, 10.5
inhomogeneous Lorentz transformations   1
inner mechanism   10.6
inner product   3.5 (3x), 5.4 (2x), 6.1, 6.2
inner product space   3.5, 6.2
insurance   3.3
integer spin   7.4
integral   3.5
integrals in quantum field field theory   7.2
interacting field theories   9.4
interactions   7.4
interpretation   8.4, 10.4
interpretation of a state   8.4
intersection   4.2
intrinsic angular momentum   2.7
invariance properties   7.4
invariant   2.6, 6.2 (2x), 7.4. 9.4
invariant state   2.5

www.ingramcontent.com/pod-product-compliance
Lightning Source LLC
Chambersburg PA
CBHW071915200326
41519CB00016B/4622